Architecting ITSM

A Reference of Configuration Items and Building Blocks for a Comprehensive IT Service Management Infrastructure

Randy A. Steinberg

Trafford rev. 01/11/2014

 www.trafford.com
North America & international
toll-free: 1 888 232 4444 (USA & Canada)
fax: 812 355 4082

Other books by Randy A. Steinberg:

Implementing ITSM
From Silos to Services—Transforming the IT Organization
to an IT Service Management Valued Partner
Trafford Press ISBN: 978-1-4907-1958-0

Measuring ITSM
Measuring, Reporting, and Modeling the IT Service
Management Metrics that Matter Most to IT Senior Executives
Trafford Press ISBN: 978-1-4907-1945-0

Servicing ITSM
A Handbook of IT Services for Service
Managers and IT Support Practitioners
Trafford Press ISBN: 978-1-4907-1956-6

If asked, Boeing Corporation can immediately show you a complete model of how an airplane is assembled right down to the smallest part . . .

If asked, NASA can immediately show you a complete model of how a space shuttle is assembled right down to the smallest nut and bolt . . .

If asked, Bath Iron Works can immediately show you a complete model of how a battleship is assembled from stem to stern . . .

If asked, can your IT organization immediately show how your infrastructure is assembled to support and deliver all your critical IT services?

If asked, can your IT organization immediately articulate everything that needs to be in place to support and deliver all your critical IT services?

IT can no longer continue to operate this way.

It's time to operate IT like a Service Organization.

—The Author

Dedication

This book is dedicated to those very hard working IT professionals and managers who deserve to see their IT solutions deploy and operate day-to-day at acceptable cost and risk to the businesses they serve.

Dedication

Table Of Contents

Book Overview 1
  Why This Book Was Written..................................... 1
  Why Have An IT Service Management Architecture?........... 4
  Architecture Analysis Tool and Other Aids........................ 6
  Book Chapters in Brief 7
Service Management Architecture.............................. 9
  Overview of the ITSM Architecture 9
  Technology Infrastructure....................................... 13
    Physical Facilities... 14
    Network.. 15
    Hardware.. 16
    Virtualization Platforms 17
    Systems Software 18
    Applications... 19
    Files and Databases 20
  Operating Infrastructure 21
    Processes 23
    Organization.. 24
    Suppliers 25
    Business Office... 26
    Project Office 27
    Services 28
    Customers... 29
  Assets and Configurations 30
  Service Transitions.. 31
  Service Designs 33
  Service Operations 34
  Strategy and Governance 36
  Knowledge 37
  Security Infrastructure.. 38
    Physical Security... 39
    Security Protocols 40
    Directory Management....................................... 40
    Identity Management 41
    Access Control... 41
    Privacy Management 42
    Application Interfaces 42
    DMZ 43

Intrusion Detection and Reporting 43
Virus and Patch Management.................................44
Auditing and Testing..44
Policy Management..45
Security Presence..45
ITSM Operating Model 46
Functional Architecture..53
Functional Architecture Overview 53
Access Management System...................................54
Automatic Call Distribution (ACD) System 55
Asset Management System 56
Auto Discovery Tool 58
Backup and Restore Management System..................60
Billing Management System....................................62
Building Management System..................................63
Capacity Management System 64
Capacity Modeling System.....................................66
Change Management System...................................68
Computer Aided Design (CAD) System 69
Configuration Management System...........................70
Console Management System 72
Customer Relationship Management (CRM) System..........73
Database Management System.................................75
Documentation Management System 76
Event Management System 77
HR Management System 79
Identity Management System...................................80
Incident Management System..................................81
Interactive Voice Response (IVR) System 83
Intrusion Detection System 84
Inventory Control System 85
IT Financial Management System.............................86
IT Service Continuity Planning System.....................87
Job Scheduling System...88
Labor Reporting System 89
Media Management System.....................................90
Network Management System 92
Performance Monitoring System...............................93
Portfolio Management System..................................94
Problem Management System 95
Process Modeling Tool..96

Procurement System..97
Project Management System 98
Prototyping System 100
Reader Board System..101
Release Build Manager......................................102
Remote Support Tool 103
Report Generator 105
Request Management System............................106
Security Management System 107
Security Test Manager 108
Service Catalog Manager 109
Service Knowledge Management System 110
Service Level Management System....................111
Software Configuration Manager 113
Software Distribution Manager...........................114
Staffing Calculator...115
Storage Management System.............................116
Surveillance System..118
System Directory..120
Test Data Generator..122
Test Management System...................................123
Uninterruptible Power Supply (UPS) System 124
Data Architecture...127
Data Architecture Overview 127
Access Management Database 128
Architecture Database.......................................128
Asset Management Database.............................129
Availability Management Database.....................130
Call Management Database................................130
Capacity Management Database........................130
Change Management Database 132
Configuration Management Database 133
Customer Relationship Management (CRM) Database 136
Definitive Media Library.....................................137
Event Management Database.............................139
Facility Management Database...........................140
Financial Management Database........................141
HR Management Database..................................142
Incident Management Database 142
IT Service Continuity Management Database 143
Knowledge Management Database.....................143

Known Error Database 144
Operations Management Database 145
Policy Database 147
Problem Management Database.................................... 148
Procurement Database 149
Project Management Database...................................... 150
Release Management Database..................................... 151
Request Management Database..................................... 152
Security Management Database..................................... 153
Service Catalog Database 155
Service Level Management Database 155
Strategy Database... 155
System Directory Database 156
Test Management Database 156
Training Database 157
Organization Architecture... 159
Organization Architecture Overview 159
Organization Models 160
Centralized Organization Model.............................. 161
Distributed Organization Model............................... 162
Networked Organization Model 163
Localized Organization Model.................................. 164
Virtualized Organization Model................................ 166
Transforming To an ITSM Organization 168
General IT Service Management Roles.......................... 176
IT Executive 177
IT Functional Unit Manager..................................... 178
Subject Matter Expert (SME) 179
Steering Group Member 180
Service Strategy Roles.. 181
IT Financial Manager 182
IT Financial Analyst.. 183
IT Financial Administrator 184
Demand Manager 185
IT Service Portfolio Analyst 186
IT Market Analyst 187
Project Manager... 188
Service Design Roles... 189
Solution Architect.. 190
Service Catalog Administrator 191
Service Level Manager 192

Service Level Analyst... 193
Availability Manager... 194
Availability Analyst.. 195
Availability Architect.. 196
Capacity Manager.. 197
Capacity Analyst... 198
Capacity Architect... 200
IT Service Continuity Manager 201
IT Service Continuity Team Leader......................... 202
IT Service Continuity Team Member 203
Chief IT Security Officer.. 204
IT Security Manager.. 205
IT Security Analyst.. 206
IT Security Auditor.. 207
Supplier Manager (Contract Manager)..................... 208
Supplier Liaison.. 209
Service Transition Roles 210
Change Manager 212
Change Administrator 213
CAB Member 214
ECAB Member 215
Change Scheduler 215
Change Owner.. 216
Release Manager.. 217
Release Owner 218
Test Manager 219
Testing and Validation Analyst 220
Configuration Manager 221
Configuration Analyst.. 222
Configuration Librarian... 223
Asset Manager.. 224
Asset Administrator 224
Procurement Analyst... 226
License Administrator.. 227
Transition Manager 228
Release Build Manager... 229
Deployment Analyst 230
Knowledge Manager 232
Knowledge Architect 233
Knowledge Administrator 234
Knowledge Owner... 235

Trainer 236
Training Administrator 237
Training Architect 238
Organizational Change Leader 240
Technical Writer 241
Service Operation Roles 242
Incident Manager 244
Incident Analyst...245
Incident Auditor 246
Service Desk Manager 247
Service Desk Analyst....................................248
Service Desk Administrator...........................249
Call Agent..250
Service Desk Infrastructure Architect 251
Request Manager...252
Request Administrator...................................253
Request Fulfillment Owner.............................254
Problem Manager 255
Problem Owner 256
Monitoring Manager 257
Monitoring Architect 258
Security Administrator...................................259
Facilities Security Administrator....................260
Physical Site Manager 261
Site Architect...262
Site Contractor .:..263
Site Technician..264
Office Manager 265
Network Operations Manager.......................266
Network Support Analyst 267
Network Technician......................................268
Network Architect...269
Network Administrator...................................270
Operations Support Manager 271
Operations Support Analyst..........................272
Operations Architect 273
Scheduler..274
Storage Administrator 275
Technical Support Manager..........................276
Technical Support Analyst 277
Systems Administrator 278

Database Administrator...279
Continual Service Improvement Roles...........................280
    Service Manager.......................................281
    Process Owner 282
    Service Owner (Product Manager) 283
    Business Relationship Manager 284
    Reporting Architect 285
    Reporting Administrator 286
    Quality Assurance Analyst.......................287
Process Architecture 289
    Process Architecture Overview.....................289
    Service Strategy Processes........................291
    Service Design Processes.........................293
    Service Transition Processes.....................295
    Service Operation Processes 297
    Continual Service Improvement Processes.........298
    Process Meta-Architecture.........................299
        Process Control 300
        Process Execution 301
        Process Enablement...........................303
    Pulling Process Architecture Together 304
Service Architecture 309
    Service Architecture Overview.....................309
    Technical Management Services 311
    Operational Management Support Services.........312
    Service Transition Support Services...............313
    Solution Design and Build Services 314
    Strategy and Control Services 315
    Hosting and Cloud Support Services...............316
    IT Business Support Services......................317
    Examples of Other Business Support Services.......318
        General Corporate Services 318
        Manufacturing Support Services 318
        Marketing Support Services 319
        Sales Back Office Support Services 319
        Sales Front Office Support Services 319
        Customer Support Services 320
        Product Support Services 320
        Procurement Support Services 320
        Educational Support Services 321
        Hospital Support Services 321

Energy and Utilities Support Services 322
Financial Trading Support Services 322
Insurance Support Services 323
Patent and Trademark Support Services 323
Banking Support Services 323
Determining Other Business Support Services 324
Assessing the Tooling Architecture 329
Tool Architecture Assessment Overview 329
Tooling Functionality Assessment............................331
Tooling Platform Assessment 332
Tooling Integration Assessment...............................333
Tooling Automation Assessment...............................334
Tooling Usability Assessment 335
Tooling Reporting Assessment 336
Tooling Data Assessment 337
Tooling Communication Assessment.........................338
Tooling Openness Assessment 339
Support Assessment...340
Mapping Tools with Processes 341
Mapping Tools with the Functional Architecture 342
Architecture Governance..343
Architecture Governance Overview 343
Architecture Governance Workflow...............................346
Architecture Exception & Appeals Workflow.................347
Architecture Vitality Workflow 348
Architecture Communications Workflow 349
About the Author 351

Chapter

1

Book Overview

Why This Book Was Written

Most people would agree that the IT infrastructure of a business is a very critical component that allows that business to operate, compete and obtain revenue. It is almost impossible to envision any business in today's world that could operate without an IT infrastructure. Yet, as critical as that infrastructure is, there is no solid blueprint that identifies all the pieces and parts of that infrastructure that need to be in place.

Let's pretend that you are a new CIO taking charge of an entire IT infrastructure. Maybe you are building this infrastructure from the ground up with a new data center. Or, maybe you wish to review the existing IT infrastructure to understand its service capabilities and resources. Where is the reference model that tells you all the pieces and parts that should be in place? What are all the data elements that are needed? Are all the right kinds of tools in place? Are all the right people and roles in place?

The purpose of this book is to attempt to close this gap by defining an enterprise IT Service Management Architecture that identifies all the types of data, tools and people resources needed to operate a successful IT Service Management infrastructure.

This book does not specify specific tools or vendor solutions to do this. Rather, a set of architecture building blocks are described that describe the types of solutions you need. So how would you use these?

- Those building a Configuration Management System (CMS) might be interested in having a comprehensive list of all the types of CIs (Configuration Items) that make up that system (See Chapter 2 for the Service Management Architecture).

- Those pulling their IT Service Management tooling strategy together may be greatly interested in using a comprehensive set of needed tool types to represent target architectures for mapping their vendor products and solutions (See Chapter 3 for the Functional Architecture plus the tools on the download website for this book).

- Those creating a Service Knowledge Management System (SKMS) might be interested in having a comprehensive set of information databases that describe all needed data elements within the Service Management architecture (See Chapter 4 for the Data Architecture).

- Those looking to identify all the needed staffing roles necessary to operate an entire service management infrastructure (See Chapter 5 for the Organization Architecture).

- Those looking to get a quick broad brush overview of all the processes needed to operate an entire service management infrastructure (See Chapter 6 for the Process Architecture).

- Those looking to get a jump start on identifying the kinds of IT and business support services with detailed descriptions that might be used for a Service catalog or Portfolio (See Chapter 7 for the Service Architecture).

- Those seeking to review and assess the value and applicability of their IT Service Management tools (See Chapter 8 to see suggested assessment criteria as well as tools on the download website for this book).

- Those seeking a general procedure for how to govern, maintain, and manage their IT Service Management Architecture (See Chapter 9 for IT Service Management Architecture Governance).

Any successful IT Service Management infrastructure needs to have a complete set of processes, data elements, organizational roles and technologies. This book presents a comprehensive set of building blocks for all of these that can serve as a target reference architecture for piecing together an entire IT Service Management infrastructure.

Why Have An IT Service Management Architecture?

IT is complex. Technologies change rapidly. On any given day, many decisions are made to implement new tools, install new services, upgrade existing infrastructures or repair infrastructure problems. How can all of this occur without a guiding blueprint or plan for how everything fits together?

A wise aviation engineer once said that there is a big difference between a million parts flying in close formation and a well-built airplane. Having an architecture for IT Service Management provides that airplane. Without it, tools are purchased for discrete needs, solutions are put into place without considering all the aspects needed to operate them successfully and important things get forgotten until an incident or a last minute scramble occurs.

Want an early quick win with this? Conduct a brief effort to map all your existing tools, management databases and organizational functions into the architectures presented in this book. Now examine the results. What is missing? How many redundant tools or people are in place doing similar things? How many tools are being paid for that provide little or no value?

Without an architecture for IT Service Management, you are essentially flying blind. You are putting your tools, processes and people together in individual pieces with no overall vision or target. This creates a high risk of redundant costs, reactive activities and missing or incomplete service management solutions.

When teams work under the assumption that they can do anything that they want, that they can use any technology specific to their immediate needs, chaos typically results. Functionality and information will be duplicated and reuse will occur sporadically if at all. Systems will not integrate well. Solutions will conflict with one another and create potential for service outages. Costs will skyrocket because similar products from different vendors, or even simply different versions of the same product, will be purchased and then operated within production. Although each individual project may be very successful, as a portfolio they may have serious challenges. It doesn't have to be this way.

An IT Service Management architecture is the organizing logic for pulling all the service management processes and IT infrastructure together. It reflects the integration and standardization requirements of your company's operating model.

Every IT Service Management process, technology solution and service asset must fit like a jigsaw puzzle piece into your future vision for your organization. The types of tools, data, processes and roles need to be understood in advance such that there is a home for every service asset and solution brought into the infrastructure. This establishes a decision framework that allows IT to articulate how the various pieces and parts of the IT infrastructure support the business and IT strategies.

Architecture Analysis Tool and Other Aids

There are a number of tools that you may find helpful related to this book. These can be downloaded at:
http://www.itsmlib.com/downloads/SitePages/Home.aspx

Use the following login:

User ID: **MeasuringITSM@itsmlib.com**
Password: **Downloads2**

The tools are:

ITSM Architectural Analysis Tool.xls

This tool can be used to assess your ITSM architecture as it exists today. You populate it with your toolsets and it identifies areas where tools may be missing, redundant or not providing any value.

CI Reference Tables.xls

This provides a simple spreadsheet that lists all the ITSM CI types, tooling types, databases and the architectural building blocks they belong to in one handy spreadsheet that can be cut and pasted into other documents or tools for your own use.

ITSM Design Guidelines (PDF Files)

This set of PDF files provides a general set of design considerations for building your ITSM Service management processes. A separate guide is provided for each phase of the ITSM Service Lifecycle.

ITSM Roles by Process Matrix.xls

This file presents a RACI chart showing each IT Service Management role by each ITSM Process. A RACI indicator is shown to indicate whether any one role is Responsible, Accountable, Consulted or Informed across all the ITSM processes in each Service Lifecycle phase.

Book Chapters in Brief

Brief descriptions of remaining book chapters are as follows:

Chapter 2—Service Management Architecture

This chapter provides a comprehensive view of the IT Service Management architecture and identifies all the recommended types of CIs (Configuration Items) that exist within it. It describes the model for how the enterprise architecture for IT Service Management is put together.

Chapter 3—Functional Architecture

This chapter provides a functional view of the IT Service Management architecture presented in the previous chapter. The functional view presented here consists of functional building blocks described as generic tool types. You can map the functional architecture shown to your existing toolsets to look for missing areas, redundant use of tools and completeness of the tools installed in your infrastructure.

Chapters 4—Data Architecture

This chapter provides a data view of the IT Service Management architecture presented in the previous chapter. The data view presented here consists of data building blocks described as generic database types. You can map your existing Service Knowledge Management System (SKMS) against this to assess completeness of your information architecture supporting your IT Service Management infrastructure.

Chapter 5—Organization Architecture

This chapter provides an organizational view of the IT Service Management architecture. In order to be effective, IT cannot just implement processes and tools. IT needs to organize itself to successfully support and deliver services. Presented here is a comprehensive set of job functions and role descriptions needed to operate an entire IT Service Management Architecture.

Chapter 6—Process Architecture

Much of what makes up the IT Service Management process architecture is written extensively in the ITSM books as well as surrounding literature. This chapter simply provides a high level overview and quick reference for all the ITSM processes.

Chapter 7—Service Architecture

This chapter provides a service view of the IT Service Management architecture. The service view presented here consists of service lines and detailed service descriptions that exist in most IT processing centers. Guidance is also provided for how to derive business support services unique to your company industry.

Chapter 8—Assessing the Tooling Architecture

This chapter provides some guidance for how to assess the value of your current IT Service Management tools.

Chapter 9—Architecture Governance

This chapter provides some guidance for how to govern and maintain your IT Service Management Architecture once it gets going.

Chapter

2

Service Management Architecture

Overview of the ITSM Architecture

This chapter provides a comprehensive view of the IT Service Management architecture and identifies all the recommended types of CIs (Configuration Items) that exist within it.

The purpose of the architecture is to provide a structure and organization around which all the myriads of technologies, processes, data and activities flow into. It provides the glue for tying the entire Service Management infrastructure together.

As you implement processes, tools, services, databases or other elements, each of these will have a home somewhere within this architecture.

By comparing your current state tools, processes and data to this architecture, you can begin to see critical infrastructure areas that may be missing. You may also find areas where redundant solutions have been deployed. Those building new services may use it as a reference point to ensure those services are adequately operational and fully supported.

The architecture presented is not technology or solution specific. Rather, it presents the building blocks and types of configuration items that need to be in place and that your specific solutions will fit into. The scope is limited to the infrastructure needed to support IT Service Management activities to support whatever services you are delivering to your business organization.

The following page illustrates a complete set of architectural building blocks suggested for an IT Service Management Architecture. These building blocks have been put into an overall framework to provide a bit of structure and order around each of the key elements.

The framework represents all the key building block areas that make up the IT Service Management architecture used to underpin IT services. Each block within the architecture can be decomposed into further subsets of building blocks. At the lowest building block level, Configuration Item Types (or Logical CIs) will be represented.

The Logical CIs represent types of infrastructure elements that should be in place for almost any infrastructure used to deliver IT services. The main goal here is to identify every logical CI needed for almost any IT service delivery and support organization.

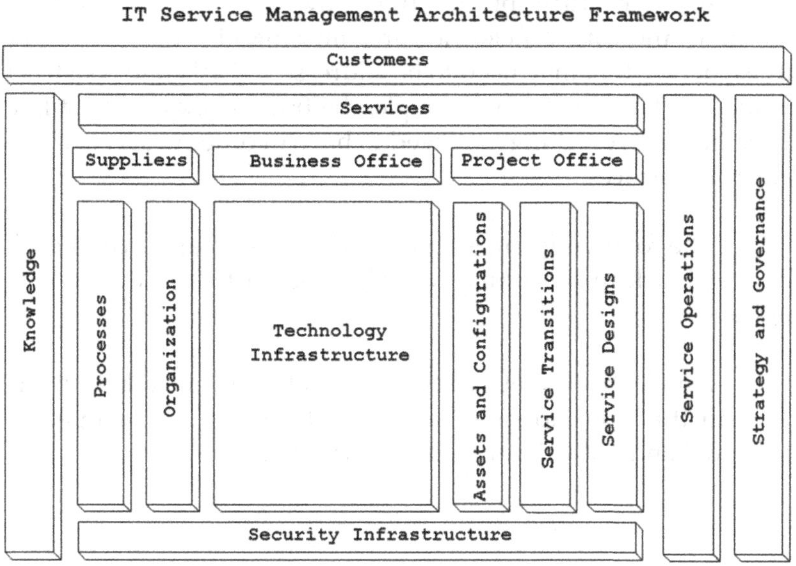

IT Service Management Architecture Framework

| Customers |
| Services |

Suppliers · Business Office · Project Office

Knowledge · Processes · Organization · Technology Infrastructure · Assets and Configurations · Service Transitions · Service Designs · Service Operations · Strategy and Governance

Security Infrastructure

Figure 1: ITSM Architecture Framework

A typical IT support and delivery infrastructure has many configuration items, technology tools, data, and work artifacts within it. The purpose of the IT Service Management Architecture Framework is to put structure and order around all the complexities needed by a successful IT delivery organization.

Each building block shown serves to provide a home for a subset of all these elements. Logical CIs, inherent within each block, can be linked to actual physical working artifacts, documents and as-built drawings within the infrastructure. This provides a schema that can be used with almost any Configuration Management tool to identify what kind of CIs should be recorded and managed.

The framework also serves another important purpose. It can be used as a definitive reference for all the architectural elements that need to be in place in order to run and operate a successful IT service delivery and support organization. Every tool, database, process and document can be mapped to it.

Infrastructure blocks that have nothing mapped to them may indicate deficiencies and missing elements in the IT infrastructure that may need to be addressed. If asked to construct a complete data center, for example, the framework is almost like a checklist of everything you need to have in order to successfully deliver IT services.

The following pages show a complete breakdown of each building block in the framework right down to the lowest level Logical CI.

An electronic set of reference tables are included on the CD with this book to show how each Logical CI is mapped to functional and data views of the architecture.

Technology Infrastructure

The Technology Infrastructure building block represents all the IT technology hardware and software resources used to support and deliver services. This includes a wide array of logical technology types such as servers, network devices, storage devices, etc.

Since there are so many of these, a lower level set of building blocks has been used to better categorize all the types of technologies that can be deployed. This lower level set of building blocks has been assembled into a secondary framework that represents a technology stack typically found in most IT organizations.

The technology stack itself is then further broken down into the Logical CIs that fall into each block. The stack itself looks like the following:

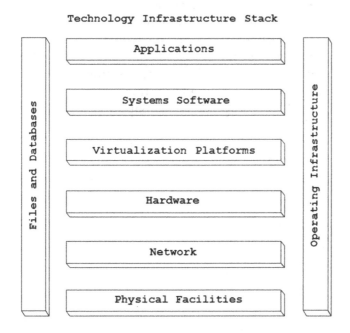

Figure 2: Technology Infrastructure Stack

Physical Facilities

Physical facilities represent actual physical locations from where IT services are delivered. It also includes all environmental CIs and other items needed to house IT equipment and people.

Physical Facilities Logical CIs are:

- Building Management Systems
- Command Center Layout Diagrams
- Cooling Chillers
- Facility Blueprints
- Facility Building Code Requirements
- Facility Cooling Diagrams
- Facility Electrical Safety Codes
- Facility Fire Safety Codes
- Facility Floor Plan Layouts
- Facility Locations
- Facility Physical Requirements
- Facility Wiring Diagrams
- Fire Suppression Systems
- Mechanical/Electrical Layouts
- Office Floor Layout Plans
- Physical Plant Equipment Documentation
- Premise Cabling Diagrams
- Uninterruptible Power Supply (UPS)

Network

Network CIs represent various network devices and equipment used to transport voice, data and video throughout the IT infrastructure.

Network Logical CIs are:

- Access Circuits (Lines)
- Domain Naming Services (DNS)
- Extranet Configurations
- Firewalls
- Hubs
- Intranet Configurations
- IP Addresses
- IP Telephony System
- Local Area Network (LAN) Configurations
- Microwave Configurations
- Modems
- Network Adapter Cards
- Network Appliance Devices
- Network Device Configurations
- Network Topology Maps
- Patch Panels
- Proxy Configurations
- Public Branch Exchange (PBX)
- Routers
- Satellite Configurations
- Sensors
- Switches
- Telephones
- Video Conference Equipment
- Virtual Private Network (VPN) Configurations
- Wide Area Network (WAN) Configurations
- Wireless Access Equipment

Hardware

Hardware CIs represent processing devices and supporting hardware used throughout the IT infrastructure.

Hardware Logical CIs are:

- Cabinets and Racks
- CD Drives
- Collators and Binding Equipment
- Consoles
- Controllers
- Copy Machines
- Desktops
- Electronic Memory Devices
- Fax Devices
- Handheld Devices
- Hardware Virtualization Configurations
- Laptops
- Mainframes
- Monitors
- Point Of Sale Devices
- Portable Disk Drives
- Printers
- RIM Devices
- Scanners
- Servers

Virtualization Platforms

Virtualization Platform CIs represent specialized hardware and software that supports sharing of IT physical resources (such as CPUs, PCs, storage and networks) among multiple applications, systems, users and files.

Virtualization Platform CIs are:

- Virtual Machine Configurations
- Virtualization Pools
- User Permissions For Virtualized Services
- Resource Share Allocations for CPU, memory, disk and network bandwidth
- Virtualized Machine IP Addresses
- Virtualized Platform Physical Locations
- Virtual Machine Operating Consoles
- Virtual Machine State Snapshots (e.g. point in time operating state records or images)

Systems Software

Systems Software CIs represent operating, middleware, file and utility systems used to provide an abstraction interface to the network and hardware devices in the IT infrastructure.

Systems Software Logical CIs are:

- Database Management Systems (DBMS)
- File Transfer Software
- Interactive Voice Response Applications
- Middleware/Messaging Software
- Network Control Systems
- Operating Systems Software
- Specialized Device Control Systems
- System Utilities
- Systems Software Inventories
- Transaction Control Software
- Virtualization Software

Applications

Application CIs represent actual application software and development tools used within the IT infrastructure.

Application Logical CIs are:

- Application Code
- Application Code Generators
- Application Code Library Configurations
- Application Development Software
- Application Programming Interfaces (APIs)
- Code Compilers
- Include and Assembly Configurations
- Software Configuration Control Systems

Files and Databases

File and Database CIs represent storage devices and logical file layout configurations used for data storage in the IT infrastructure.

File and Database Logical CIs are:

- Channel Subsystems
- Data Architecture
- Data Dictionaries
- Data Record Layouts
- Databases
- Disk/RAID Configurations
- Files
- IO Subsystems (IOS or BIOS)
- Logical Data Models
- Storage Adapter Cards
- Storage Area Networks (SANs)
- Storage Compression Schemes
- Storage Configuration Layouts
- Storage Devices
- Storage Management Systems

Operating Infrastructure

Operating Infrastructure CIs represent software, systems and configurations used to provide a centralized control point for managing events, incidents, problems, changes, operational activities and reporting throughout the IT infrastructure.

Operations Management Logical CIs are:

- Asset Management Software
- Auto Discovery Systems
- Automatic Call Distribution Systems (ACD)
- Backup and Restore Tools
- Call Management Software
- Change Management Software
- Chargeback Software
- Checkpoint Restart and Control Software
- Command Center Configurations
- Configuration Management Software
- Definitive Software Libraries
- Distribution List Management Software
- Documentation Support Software
- Event Correlation Schemes
- Event Monitoring Software
- File Transfer Schedules
- Hardware Maintenance Mgt Software
- HW/SW/NW Maintenance Procedures
- Incident Management Software
- IT Service Continuity Planning Systems
- Job Run Schedules
- Media Archive Control Software
- Media Management Software
- Monitoring Agents
- Operational Run Books
- Operational Scripts
- Patch Management Support Software
- Performance Monitoring Software
- Problem Management Software
- Reader Displays
- Remote Support Systems

- Service Catalog Systems
- Service Desk Software
- Service Reporting Systems
- Shift Turnover Reports
- Software Distribution Managers
- Software Maintenance Mgt Software
- Supply and Inventory Control Systems
- System Consoles
- System Directories
- Tape Management Software
- Time Reports
- Toll-Free Access Lines
- Transaction Control System Configurations

Processes

Process CIs represent structured sequences of activities, procedures and workflows used to manage and operate the IT infrastructure.

Process Logical CIs are:

- Activities
- Procedures
- Process Descriptions
- Process Architecture
- Process Guides
- Process Metrics
- Process Modeling Tools
- Process Models
- Process Quality Reports
- Work Instructions

Organization

Organization CIs represent roles, jobs and skills that are used to describe people resources used to operate, manage, support and deliver IT services.

Organization Logical CIs are:

- Job Descriptions
- Organization Architecture
- Organization Chart
- Role Descriptions
- Roles and Responsibilities Matrix
- Skills Inventory

Suppliers

Supplier CIs represent third party vendors used to support services delivered by the IT infrastructure.

Supplier Logical CIs are:

- Supplier Catalogs and Brochures
- Supplier Contact Lists
- Supplier Quality Metrics
- Supplier Quality Reports
- Supplier Cost Estimates
- Supplier Descriptions

Business Office

Business Office CIs represent work artifacts used to manage financial resources and transactions as well as people resources used to support and deliver IT services.

Business Office Logical CIs are:

- Account Code Lists and Descriptions
- Budget and Expenditure Forecasts
- Budget Line Item Descriptions
- Budget Reports
- Charging Algorithms and Calculations
- Charging Code Lists and Descriptions
- Charging Summary Report
- Cost Estimates
- Cost Models
- Cost Pool Descriptions
- Escalation Contact Lists
- HR Recruiting Logs
- Network Carrier Service Agreements
- Office and Supply Purchase Receipts
- Personnel Records
- Purchase Orders
- Service Catalog
- Service Charging Invoices
- Service Proposals
- Software License Obligations
- Software Licenses
- Staffing Estimates and Projections
- Supplier Contact Lists
- Telephone Numbers
- Underpinning Contract (UC)
- Vendor Catalogs
- Vendor Invoices
- Vendor Leases
- Work Orders

Project Office

Project Office CIs represent work artifacts used to manage projects and programs used to design, transition, and improve the IT infrastructure and the services it supports.

Project Office Logical CIs are:

- Project Earned Value Analysis Reports
- Project Issue Lists
- Project Portfolio
- Project Status Reports
- Project Templates
- Project Work Plans
- Project Workbooks
- Project Working Standards
- Scope Statements
- Service Improvement Plans (SIPs)
- Work Breakdown Structures

Services

Service CIs represent work artifacts used to describe, measure, report and market IT services.

Service Logical CIs are:

- Operational Level Agreement (OLA)
- Service Brochures
- Service Compliance Requirements
- Service Dashboard
- Service Descriptions
- Service Level Agreement (SLA)
- Service Marketing Plans
- Service Metrics
- Service Quality Reports
- Service Quality Survey Results
- Service Quality Surveys
- Service Reports
- Service Review Meeting Notes

Customers

Customer CIs represent work artifacts used to manage relationships with customers and identify their needs and values.

Customer Logical CIs are:

- Business Continuity Plans
- Customer Constraints
- Customer Contact History
- Customer Requirements
- Customer Sales Calls
- Customer Satisfaction Surveys
- Customer Satisfaction Survey Results
- Relationship Contact Listings
- Relationship Issues and Status
- Vital Business Functions

Assets and Configurations

Asset and Configuration CIs represent inventories, configurations, descriptions and models for what is in the IT infrastructure and how all the pieces and parts fit together to provide IT services.

Asset and Configuration Logical CIs are:

- Asset Configurations
- Asset Inventory Reports
- Asset Listings
- Asset Records
- Asset Tags
- Circuit Numbers
- Configuration Audit Results
- Configuration Databases
- Configuration Item Naming Conventions
- Configuration Item Records
- Configuration Reports
- Equipment IP Addresses
- Equipment Physical Configuration Specifications
- IP Addressing Schemes and Configurations
- IP/MAC Addresses and Owners
- Packing Slips
- Port Assignments
- Service Models

Service Transitions

Service Transition CIs represent work artifacts and elements of the infrastructure used to support of build and transition activities to move new or changed IT services to a production operational state.

Service Transition Logical CIs are:

- CAB Meeting Minutes
- CAB/ECAB Member Listings/Distribution Lists
- Change Approval Packages
- Change Historical Reports
- Change Status Reports
- Change Tickets
- Definitive Hardware Store Inventory (DHS)
- Definitive Media Library (DML) Inventory
- Deployment Schedules and Plans
- Device Configuration Images
- Equipment Warranty Records
- Forward Schedule of Changes (FSC)
- IT Service Continuity Plan
- IT Service Continuity Test Plan
- IT Service Continuity Test Results
- IT Service Continuity Testing Schedules
- Management Information Base (MIB)
- Migration/Deployment Plans
- Planned Service Availability (PSA) Notices
- Post Implementation Review (PIR) Findings
- Production Readiness Review Checklists
- Release Packages
- Requests For Change (RFCs)
- Service Test Plans
- Site Implementation Plans
- Site Surveys
- Software License Keys
- Solution Review Findings
- Test Data Files

- Test Lab Layouts and Configurations
- Test Plan
- Test Results
- Testing Schedules
- Use Cases

Service Designs

Service Design CIs represent work artifacts and elements of the infrastructure used to design new or changed IT services to meet customer utility and warranty requirements.

Service Design Logical CIs are:

- Capacity Data Collection Applications and Scripts
- Application and Service Sizing Estimates
- Availability Database
- Availability Plans
- Availability Report
- Business Forecasts
- Capacity Baselines
- Capacity Data Analytics
- Capacity Database
- Capacity Models
- Capacity Plan
- Capacity Report
- Capacity Thresholds
- Demand Factor Descriptions
- Demand Forecasts
- Design Principles and Guidelines
- Disk Utilization Reports
- Engineering Specifications
- Event Monitoring Plan/Architecture
- Hardware Maintenance Requirements
- Network Traffic Reports
- Resource Forecasts
- Security Vulnerability Assessment
- Server Utilization Reports
- Service Design Packages
- Service Solution Prototypes and Demos
- Software Maintenance Requirements
- Storage Usage Reports
- System Performance and Utilization Reports
- Vital Business Function (VBF) Inventory
- Workload Characterizations

Service Operations

Service Operation CIs represent work artifacts and elements of the infrastructure used to operate and deliver services on a day to day basis.

Service Operation Logical CIs are:

- Backup Schedules
- Badge Control Logs
- Building Management System Event Logs
- Call Handling Report
- CCTV Security Camera Video Archives
- Customer Contact Lists
- Customer Statements For Major Incidents
- Event Definitions
- Event Descriptions
- Event Filtering Criteria Lists
- Event Handling/Escalation Lists
- Event Logs
- Event Status Reports
- Event/Alarm Thresholds
- File Transfer Verification Logs
- Firewall Logs
- Incident Classification Schema
- Incident Notification Lists
- Incident Status Reports
- Incident Tickets
- Intrusion Detection Log
- Job Control Logs
- Job Schedule
- Job/Script Control Data (JCL)
- Known Errors
- Maintenance Records
- Offsite Storage Archival Pick Lists
- Operational Plan
- Operational Run Books
- Operational Run Logs
- Operational Scripts
- Operations Staffing Plan

- Physical Building Security Event Logs
- Problem Status Reports
- Problem Tickets
- Process Owner Contact List
- Repair Dispatch Tickets
- Repair Service Schedules
- Report Distribution Schedules
- Repot Distribution Lists
- Request Fulfillment Procedures
- Request Status Reports
- Request Tickets
- Resource Startup/Shutdown Procedures
- Service Desk Call Handling Statistics
- Service Owner Contact List
- Service Requests
- Service Startup/Shutdown Procedures
- Shift Turnover Reports
- Spare Parts Inventories
- Staffing Schedules
- Supply Inventories
- Supply Inventory Counts
- Supply Inventory Lists
- Supply Reorder Points
- System Performance and Utilization Data
- Tape and Media Archives
- Tape Vault List
- UPS Startup/Shutdown Procedures
- Visitor Access Control Logs

Strategy and Governance

Strategy and Governance CIs represent work artifacts used to provide strategies for services and govern their delivery.

Strategy and Governance Logical CIs are:

- Architecture Standards
- Change Policy
- Clock/Time Management Policy
- Configuration Control Policy
- Data Retention Policy
- Demand Strategies
- Documentation Templates
- Enterprise Architecture
- HR Policies
- Incident Escalation Policy
- Incident Prioritization Policy
- Intellectual Property (ICAP) Policies
- IT Research
- IT Service Continuity Invocation Policy
- Management Tooling Architecture
- Market Research
- Naming Standards
- Network Architecture
- Patch Management Policy
- Quality Control Plan
- Release Policy
- Risk Assessment
- Risk Register
- Security Incident Response Plan
- Security Policy
- Service Portfolio
- Technical Architecture
- Tooling Requirements

Knowledge

Knowledge CIs represent work artifacts and repositories used for gathering, analyzing, storing and sharing knowledge and information used to operate the IT infrastructure, support and deliver IT services.

Knowledge Logical CIs are:

- Business and IT Notices and Announcements
- Business Presentations
- Communications Plan
- Customer Communication Notices
- Documentation Subscription Lists
- Facility Emergency Exit Procedures
- Governance Strategy
- Hardware and Software Benchmarks
- Holiday Schedules
- How-To Help Knowledge
- IT Audit Results
- Known Errors
- Management and Line Staff Presentations
- Meeting Agendas
- Meeting Minutes
- Physical Equipment Maintenance Manuals
- Research Reports
- Research Subscriptions
- Security Awareness Plan
- Service Sourcing Strategy
- Training Guides
- Training Materials
- Training Plans
- Training Schedules
- Training Setup Requirements
- Vendor Manuals
- Whitepapers and Articles of Interest

Security Infrastructure

The Security Infrastructure building block represents all the IT technology resources and capabilities used to secure and protect service assets in the IT infrastructure. This infrastructure is used to protect the confidentiality, integrity and availability of services from unauthorized access or use.

Security infrastructures and activities cover a wide array of solution and services to protect the infrastructure. For this reason, a lower level set of building blocks has been used to better categorize security related CIs. This lower level set of building blocks has been assembled into a secondary framework that represents a complete set of security facility and operational elements that need to be in place to support almost any IT infrastructure.

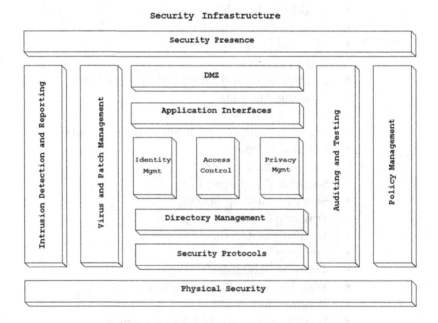

Figure 3: Security Infrastructure CIs

Physical Security

Physical Security CIs represent artifacts used to secure processing sites from unauthorized physical access.

Physical Security Logical CIs are:

- Badges
- Biometric Devices
- Door Lock Combinations
- Key Lists
- Security Keypads
- Security Scanning Devices
- Surveillance Logs
- Video Surveillance Devices

Security Protocols

Security Protocol CIs represent secure protocols used to block access to data from unauthorized applications as it is transmitted over a network.

Security Protocol Logical CIs are:

- Encryption Schemes
- Internet Protocol Security (IPSec) Controls
- Secure Electronic Transmission (SET) Controls
- Secure Hypertext (S-HTTP, HTTPS) Controls
- Secure Shell (SSH) Controls
- Secured Transmission Logs
- Secure Socket Layer (SSL) Controls
- Virtual Private Network Configurations (VPNs)
- SFTP Secure File Transfer Protocol
- TLS Transport Layer Security

Directory Management

Directory Management CIs represent system directories, their structures, permissions, interfaces and management software.

Directory Management Logical CIs are:

- Directory Interfaces
- Directory Management Software
- Directory Names
- Directory Permissions
- Directory Objects and Structures

Identity Management

Identity Management CIs represent systems and interfaces that restrict access to services and service assets via user ID and password protection.

Identity Management Logical CIs are:

- Identity Management Applications
- Identity Management Interfaces
- Passwords
- Security Profiles (Matching Patterns)
- Single Sign-On (Password Sync) Systems
- User IDs
- Security Access Markup Language (SAML) Controls and Configurations

Access Control

Access Control CIs represent security profiles and access control lists to ensure only authorized users gain access to services and service assets in the infrastructure. This category also includes visitor logs to control physical access to processing sites and facilities.

Access Control Logical CIs are:

- Access Control Lists
- Access Permissions
- Access Profiles
- Visitor Control Logs

Privacy Management

Privacy Management CIs represent policies and systems that enforce IT security policies to ensure that services and service assets, especially data, are accessed only by those allowed to view them.

Privacy Management Logical CIs are:

- Policy Enforcement Systems
- Privacy Configurations
- Privacy Policies

Application Interfaces

Application Interface CIs represent security applications and interfaces accessed by production services, usually their supporting applications, to control or gain access to protected services and service assets.

Application Interface Logical CIs are:

- Security Application Interfaces
- Security Application Protocols
- Application IDs and Control Lists
- Security Applications

DMZ

DMZ CIs represent service assets that make up the "Demilitarized Zone" that provides a buffer around the IT infrastructure to protect it from hostile attacks and intrusions launched over the internet.

DMZ Logical CIs are:

- Firewall Appliance Devices
- Firewall Configurations
- Firewall Rules
- Firewall Servers
- Proxy Configurations
- Proxy Servers

Intrusion Detection and Reporting

Intrusion Detection and Reporting CIs represent security monitoring and reporting systems used to protect the IT infrastructure from hostile attacks.

Intrusion Detection and Reporting Logical CIs are:

- Host-based Intrusion Monitors
- Intrusion Detection Reports
- IP Packet Scanners
- Network-based Intrusion Monitors
- Password Cracking Monitors
- Security Incident Reports
- Security Logs
- Unauthorized Access Reports

Virus and Patch Management

Virus and Patch Management CIs represent security patches, communications and anti-virus updates used to proactively detect and prevent exploitation of security vulnerabilities in the infrastructure.

Virus and Patch Management Logical CIs are:

- Anti-Virus Patches and Downloads
- Anti-Virus Subscription Lists
- Forensics
- Security Alert Notices
- Security Patches

Auditing and Testing

Virus and Patch Management CIs represent security patches, communications and anti-virus updates used to proactively detect and prevent exploitation of security vulnerabilities in the infrastructure.

Auditing and Testing Logical CIs are:

- Denial of Service Testing Software
- Intrusion Testing Software
- Security Audit Control Requirements
- Security Audit Results
- Security Test Conditions
- Security Test Results
- Vulnerability Testing Software

Policy Management

Policy Management CIs represent security policies and requirements that govern the security solutions used to protect the IT infrastructure.

Policy Management Logical CIs are:

- Information and Document Classifications
- Security Escalation Contact Lists
- Security Policies
- Security Requirements

Security Presence

Security Presence CIs represent overall presence of the business organization across the security infrastructure. This also includes external security interfaces to trusted partners, other business organizations and suppliers.

Security Presence Logical CIs are:

- External Security Interfaces
- Federated Security Configurations
- Trusted Partners

ITSM Operating Model

Experience with many IT organizations has shown that there is still much confusion over basic IT operating concepts. Many IT personnel will get totally confused over items such as:

What is a request versus a change?
What is a release versus a change?
Is this a project or a request?
Is this a service or a request?
What is a service?
What is a request?
What is a role versus a function?
Is a role the same thing as a job?
Is that user asking for a change or a request?

The ITSM Operating Model on the following page has been identified to overcome much of this confusion. Its intent is to put each key operating element (requests, incidents, changes, etc.) into an overall context that can be directly understood.

The operating model itself is explained further on succeeding pages.

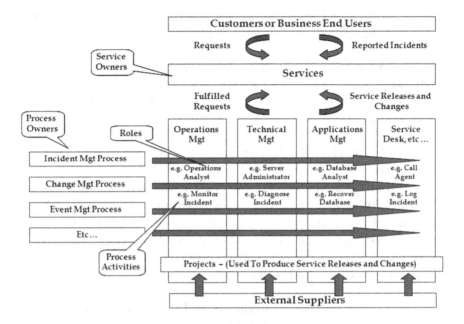

Figure 4: Suggested ITSM Operating Model

The operating model is based on the following basic tenets that describe how IT delivers its services:

- Customers get services;
- Services are delivered through one or more functional units;
- Functional units deliver their services through consistent and repeatable processes
- IT Service Management cannot take place without service and process ownership in the organization

The model itself can best be described through a number of operating principles described as follows:

A Customer Is Anyone Who Receives a Service. Anyone who gets an IT service is viewed as a customer. The customer perception of value will ALWAYS determine whether the service is of value.

A Service Is Anything of Value Delivered To a Customer. The customer does not care about anything else IT does other than whether the intended value from the service is being delivered. A NOC (Network Operations Center) is not a service—it is a functional organizational unit. An application is not a service—it is part of the many service assets used to deliver a service. Products (i.e. software, utilities, hardware, etc.) are not services—they are also service assets.

Customers Only Interact With IT in Two Ways beyond Receiving Their Services: They Issue Requests or Report Incidents. Outside of receiving their services, customers will only interact with IT through requests and reporting of incidents (disruptions to their services).

Customers Always Issue Requests—Never Changes. Customers will NEVER issue IT changes. They may submit a request for something. IT may translate that request into an IT change within their operating model in order to fulfill it.

A Request Is a Transaction against a Service. A request is anything asked from IT that is not an incident. It is a transaction that is fulfilled by IT against one more services that are being provided. Attempts to describe requests as anything else (i.e. as services, standard changes, "requestable" services, non-project requests, etc.) only confuse everyone and invite trouble.

Large Complex Needs and Activities (i.e. "Move a Data Center") are Still Requests. Larger requests like these should always be treated as requests to ensure they are properly addressed. The above example should really be broken down into smaller requests such as "Provide an Estimate for Moving the Data Center" or, "Produce a Project Charter for Moving the Data Center".

IT Delivers Services through Functions. Services are delivered through functional organization units. Examples of these include Technical Management, Operations Management, Applications Management, Service Desk, and the Network Operations Center (NOC).

Suppliers Support IT Functions in Their Outcomes through Goods and Services. Suppliers provide goods (hardware, software, etc.) and services (maintenance, support, etc.) to IT. Even if many services are outsourced to suppliers, IT is still ultimately responsible for the quality of the services being delivered.

Functions Perform Their Tasks Guided By Consistent Repeatable Processes. IT Functions execute their tasks using consistent processes that transcend one or more functional units. This ensures that an expected and consistent level of IT quality is being achieved. Variation in process has been shown to consistently lead to unexpected errors, incidents, wasted labor, service outages and higher operating costs.

Everything IT Implements Is Guided By Changes and Implemented Through Releases. Change Management is about assuring that the right steps and approvals have taken place to assure a change can be implemented with minimal risk and disruption to services. Release Management is the physical act of implementing one or more changes. Clear separation of duties should occur wherever possible. Avoid getting in the trap of calling some changes "Changes" and other kinds of changes "Releases".

Projects Exist As a Means to Enhance or Create New Service Solutions. Projects are not services, nor are they requests. They exist as logical units of work to provide an intended outcome or objective. They typically result from initiated requests to add, change or remove services.

Service Owners and Process Owners Form the Organizational Foundation for IT Service Management. At a very minimum, Services should be identified and accountability for their delivery and quality assigned through the Service Owner role. Process accountability is assigned through the Process Owner role. Without these roles in existence, there is no accountability for services or processes in the IT organization—a basic organizational requirement to operate as an IT service provider.

Roles Are Not the Same as Job Positions. Roles are virtual groupings of process and service activities. These are then assigned to job positions that physically exist in the company's IT organization chart. A job may consist of a single role, multiple roles or only part of a role.

Applications Are Simply Another Form Of Service Asset. Applications should NOT be viewed differently from operations in an IT Service Management delivery organization. They are service assets that when included with other service assets, form the structure and dynamics for a successful service solution. Both applications and their supporting operational infrastructure need to go through all the ITSM Service Lifecycle stages together.

Chapter

3

Functional Architecture

Functional Architecture Overview

This chapter provides a functional view of the IT Service Management architecture presented in the previous chapter. The functional view presented here consists of functional building blocks described as generic tool types.

Each generic tool type is listed with a high level description and a set of functional attributes. These can loosely be translated into tool functional specifications for building or procuring IT Service Management tools.

The functional architecture can also serve as a reference set to compare your installed tools against. Through this comparison, you can assess whether your existing tool set is complete, has missing elements or is redundant in various areas. The download website with this book has a Technology Assessment Tool which can be used to assess the value your tools provide. It also provides a Tool Heat Map to assess for architecture completeness in your toolsets.

Access Management System

This automates handling and administration of user profiles, access control lists and authentication to allow access to IT resources.

- Store, manage and maintain user Ids and passwords
- Prevent unauthorized access by using a security policy server to enforce security across multiple file types, application providers, devices and protocols
- Load balancing to prevent performance bottlenecks
- Interfaces with advanced authentication technologies such as retina scanning or biometrics
- Stores profiles for advanced authentication technologies such as retina scanning or biometrics
- Rules based authorization engine
- Multiple directory support
- Customer Self-Registration Template
- Interface to Security Management System
- Provide reports that support security access audits

Automatic Call Distribution (ACD) System

This automates handling and response of telephony calls to minimize human intervention. Directs calls to appropriate parties automatically. Provides statistics on calls received and processed.

- Process calls on a first-come first-serve basis
- Answer all calls immediately
- Place callers in call-queue and direct to next available Call Agent
- Support multiple call queues
- Log call group activity
- Provide capability to monitor call group activity, analyze call queues and Call Agent hold time
- Provide different call processing paths based on defined business rules
- Provide capability for calling route priority handling
- Allow for establishment of thresholds to minimize delays and route calls to alternative queues
- Allow for routing of calls based on incoming phone number dialed (DNIS)
- Provide incoming line capacity adequate to support the desired infrastructure
- Interface with corporate telephone and communications systems
- Manage incoming call traffic by employing a queuing or waiting list assignment to each call
- Provides management information reports on agent performance
- Provides real time monitoring of workload
- Provide capability to route to a recorded message and then connect to the first available agent if all agents are busy
- Provide logging capability for call management statistics
- Provide interfaces to Incident and Financial Management databases

Asset Management System

This automates business processes and controls that join financial, contractual and inventory functions for physical IT assets such as hardware and software.

- Storage of information about infrastructure hardware, software and network assets
- Linkage to Configuration Management databases
- Support for Asset status and lifecycle management such as procurement, stored, configured, deployed, active and retired stages to support release impact analysis, planning, rollout and deployment activities
- Ability to track a wide variety of enterprise IT asset types (hardware, software and services)
- Ability to record a wide variety of contracts and licensing agreements
- Provides direct support for software license management
- Ability to track the physical location of contracts and agreements, and identify the individuals responsible for them
- Ability to interface data with and from automated discovery applications
- Ability to group an individual customer's/user's assets and services to provide cost information
- Ability to generate reports based on geography, user, CI type and impact
- Ability to perform ad hoc/general queries
- Ability to integrate with other ITSM databases and applications
- Ability to manage leases, depreciation schedules, warranties, and service provider contracts as critical elements
- Ability to track both fixed and variable costs of IT assets
- Ability to tie IT assets to related items (e.g., individuals, contracts, leases, warranties, service-level agreements, other assets, incidents, problems)
- Integration with ITSM Configuration Management Database

- Ability to support a web-based front end that can significantly reduce the amount of administrative overhead
- Ability to manage moderate-to-large quantities of complex IT asset data using a standard, non-proprietary database engine
- Ability to support both flexible data import/export, and simple points of integration for associated tools
- Interfaces with Auto Discovery and Inventory Control tools to automate gathering of asset and inventory information.

Auto Discovery Tool

This automates gathering of information and configurations of IT assets such as hardware, software and networking devices that are reachable through the network.

- Automated checks for sufficient available disk space, hardware and software prerequisites in the live environment to ensure compatibility for new or changed releases
- Hardware and software audit reporting
- Flexible ability to filter, add or limit audit items being checked for or to match against a defined set of configuration requirements
- Automatic event notification for workstations or servers that fail audits
- Hardware and software audit reporting
- Flexible ability to filter, add or limit audit items being checked for or to match against a defined set of configuration requirements
- Allow for collection of data from diverse sources including systems, networks, applications, web servers and phone systems
- Provide a Data Warehouse capability for storage of availability related information
- Provide storage of historical availability plans, risk assessments and reports
- Data can be archived to tape and restored as needed
- Provide interfaces to ITSM repositories such as Incident, Problem, Service Level, Capacity and Financial databases
- Provide interfaces to System and Security Monitors and Event Correlation Tools
- Components may be customized to meet specific requirements related to the extraction, integration, storage and management of system availability from multiple sources
- Data access extends across multiple hardware platforms and operating systems, networks, web servers, telephone

systems, and includes statistics generated by custom applications
- Interfaces with Report Generation Tool
- Interfaces with Asset, Configuration and Inventory Control tools

Backup and Restore Management System

Automates scheduling, handling and processing of backup and restore activities. This also includes automated handling of backup media between live, archive and offsite locations.

- Ability to backup / restore files locally and over the network
- Ability to backup network operating system specific files
- Ability to backup system files
- Ability to communicate to user status of file, location of data for recovery
- Ability to do both full and incremental backups
- Ability to generate archive cycle list
- Ability to generate recovery scripts
- Ability to handle data compression
- Ability to identify logical groupings of files
- Ability to manage backup media
- Ability to operate across different platforms
- Ability to perform open file backups
- Ability to point in time restores
- Ability to provide problem notification
- Ability to restore alternate path
- Ability to restore archived data
- Ability to restore back to specific time
- Ability to restore by logical groupings of files
- Ability to restore cross platform
- Ability to restore from incremental and full backups
- Ability to restore individual files from full backup
- Ability to restore system files
- Ability to support and modify retention rules
- Ability to support multiple media
- Ability to verify success / failure of backup/restore activities
- Access to an archive and historical database
- Access to central repository for archive / historical database
- Catalog of media location / status

- Provide command line and GUI interfaces
- Support high volumes in short periods of time
- Search engine to find and call up media
- Support for hot backups

Billing Management System

This automates billing activities for IT services. Includes calculation of billing charges, submission of bills to customers, revenue validation and interfaces these activities with IT financial systems.

- Integrate with Capacity Management Database for usage information
- High flexibility provided for development of charging algorithms based on complex business logic
- Can allocate IT resources to specific cost centers
- Able to apply unique rates to a variety of IT services
- Support various billing modes such as zero-balancing, pro-ration and others
- Provide support for corporate audit functions
- Support various invoicing calendars
- Provide multi-currency support
- Provide capability to mix charge-back methodologies, and to easily apply changes in response to new requirements
- Provide capability for business units to track, analyze and manage their own usage and costs
- Effectively support and manage billing and invoicing cycles
- Interfaces with IT Financial Management systems

Building Management System

This automates monitoring and alerting of physical building environmental and safety concerns such as electrical, heating, cooling, lighting and fire suppression facilities.

- Provide high technology system installed on buildings housing processing, network equipment and data centers that controls and monitors the building's mechanical and electrical equipment such as air handling and cooling plant systems, lighting, power systems, fire systems, and security systems.
- Actively monitor all major critical facility components and fire alarm panels for alarms and notifications.
- Provide and maintain a notification hierarchy process to alert all critical engineering specialists and security personnel of facility alarms and alerts.
- Provide an Element Management Systems (EMS) for all infrastructures with SNMP protocols with loss of service alerts that can be displayed on systems management consoles.
- Open an incident ticket in the Incident Management system to report and track the status of any facilities related incident.
- Utilize a Security Monitoring System to monitor and manage all infrastructures with SNMP protocols.
- Provide video surveillance and critical security data to a security management console.

Capacity Management System

This automates processing of IT resource performance and usage data to indicate whether enough resources are in place to adequately support services. It includes analysis of performance data to indicate capacity trends and reports on capacity usage and levels.

- Allow for collection of data from diverse sources including systems, networks, applications, web servers and phone systems
- Provide a Performance Data Warehouse capability
- Provide storage of historical capacity models and reports
- Provide capability to automatically reduce data over time to conserve disk resources, while enough data is retained for baseline, trend and exception analysis
- Data can be archived to tape and restored as needed
- Provide interfaces to ITSM repositories such as Incident, Problem and Financial databases
- Components may be customized to meet specific requirements related to the extraction, integration, storage and management of system performance data
- Data access extends across multiple hardware platforms and operating systems, networks, web servers, telephone systems, and includes statistics generated by custom applications
- Interface with statistical calculation and summarization tools (i.e. SAS, SPSS, etc.)
- Analyzes performance and operational data from different CI types
- Analyzes service chains of CIs to obtain end-to-end performance
- Interfaces to Service Level Management tools to track performance against established service targets
- Provides real-time and historical analysis of performance data and trends
- Allows for drill-down capabilities to users, CIs and processes active at the time of a performance problem
- Provides support for exploration of cause-and-effect relationships

- Uncovers cycles and patterns in system behavior
- Provides alerting mechanisms to notify staff of potential problems before users are affected
- Provides real-time views with up-to-the-minute performance data
- Interfaces to Capacity Management Database to maintain a historical record of system and application performance
- Identifies trends, cycles and patterns to better understand performance
- Provides trend analysis capabilities to prepare for and avoid performance problems and outages
- Provides interfaces to application programs and developers to allow for performance monitoring within application segments and code
- Provides interfaces to database systems to monitor performance of those systems as well as access calls to them

Capacity Modeling System

Automates construction, management and execution of capacity models used to predict and forecast IT resource usage against future business demand.

- Predicts process, transaction and job response times based on system loads and volumes
- Predicts hardware utilization levels based on system loads and volumes
- Allows for correlation between specified business workloads and volumes with system workloads and volumes
- Allows for what-if scenario modeling including changes in hardware and software platforms as well as workload volumes
- Predicts application performance under varying loads
- Allows for prediction for unexpected spikes in demand
- Identifies performance bottlenecks within the infrastructure being modeled
- Provides workload and queue result statistics both summarized and in detail
- Interfaces to reporting tools for creation of capacity forecasts
- Provides flexibility in modeling such as standard M/M/1 queuing or stochastic modeling approaches
- Allows for storage of multiple versions of models
- Provides capabilities to create baseline models and compare results to existing performance
- Provides capabilities to compare what-if scenarios to baseline models
- Provides flexible resource modeling components based on actual vendor platforms
- Allows for storage of multiple versions of models
- Provides capabilities to create baseline models and compare results to existing performance
- Provides capabilities to compare what-if scenarios to baseline models
- Provides flexible resource modeling components based on actual vendor platforms

- Capability to model and simulate application solutions before implementing them in a production environment reducing costly and time consuming test cycles
- Capability to evaluate application flow and logic, discover and remove bottlenecks, and predict needed performance and capacity
- Capability to track state and logic of applications as they execute in test environment to identify potential performance bottlenecks
- Logging function to track recorded performance and timing information
- Ability to simulate transaction volumes
- Capability to model complex changes in application path based on business logic or other rules

Change Management System

This automates the workflow for logging, processing, approving and reporting on changes in the IT infrastructure.

- Repository to store, retrieve and report on RFCs in an easily accessible format
- Capability to link RFCs to projects
- Links to the Configuration Management database or Configuration Management data
- Means to identify easily the other CIs that will be impacted whenever a Change to any specific CI is proposed
- Ability to allow Change Management staff, Change builders, testers, etc. to add text to Change records
- Ability to store back-out procedures should a Change cause problems
- Ability to store historical reports on Change Management activity
- Automatic production of requests for impact and resources assessment to the owners of the impacted CIs
- Ability for all authorized personnel to submit RFCs from their own terminal or location
- Ability to progress requests through the appropriate stages of authorization and implementation and to maintain clear records of this progress
- Automatic warnings of any RFCs that exceed pre-specified time periods during any stage
- Automatic prompting to carry out reviews of implemented Changes
- Ability to build Changes
- Automatic production of FSCs
- Process/workflow feature
- Ability to link RFCs to forward change events and Forward Schedule of Change (FSC)
- Calendaring capabilities that can be viewed by those requesting changes as well as those administering and implementing changes
- Ability to easily reschedule changes and identify scheduling conflicts

Computer Aided Design (CAD) System

This automates detailed engineering, design and drawings of physical components for conceptual design and layout. It includes 3D and 2D dimensional models and drawings.

- Support development of engineering as built design documents, blueprints, construction layouts, engineering documentation, manufacturing drawings, and Bills of Materials
- Provide for wireframe geometry creation
- Provide for 3D parametric feature based modeling
- Provide Solid and Freeform surface modeling
- Automate design of assemblies, which are collections of parts and/or other assemblies
- Create engineering drawings from the solid models
- Reuse design components
- Allow for modification of design of models and the production of multiple versions
- Automate generation of standard components for designs
- Validate and verify designs against specifications and design rules
- Simulate designs without building a physical prototype
- Provide import/export interfaces to exchange data with other software packages
- Maintain libraries of parts and assemblies
- Calculate mass properties of parts and assemblies
- Aid visualization with shading, rotating, hidden line removal, etc.
- Synchronize modifications of any feature reflected in all information relying on that feature; drawings, mass properties, assemblies, etc.

Configuration Management System

Automates the storage, management, maintenance and reporting of IT configuration records and the relationships between configuration items.

- Security controls to limit access on a need-to-know basis
- Support for CIs of varying complexity e.g. entire systems, Releases, single hardware items, software modules, or hierarchic and networked relationships between CIs; by holding information on the relationships between CIs, Configuration Management tools facilitate the impact assessment of RFCs
- Easy addition of new CIs and deletion of old CIs
- Automatic validation of input data (e.g. are all CI names unique)
- Automatic establishment of all relationships that can be automatically established, when new CIs are added
- Support for CIs with different model numbers, version numbers, and copy numbers
- Automatic identification of other affected CIs when any CI is the subject of an Incident report/record, Problem record, Known Error Record or RFC
- Integration of Problem Management data within the CMDB, or at least an interface from the Configuration Management system to any separate Problem Management databases that may exist
- Automatic updating and recording of the version number of a CI if the version number of any component CI is changed
- Maintenance of a history of all CIs (both a historical record of the current version—such as installation date, records of Changes, previous locations, etc.—and of previous versions)
- Support for the management and use of configuration baselines (corresponding to definitive copies, versions etc.), including support for reversion to trusted versions
- Ease of interrogation of the CMDB and good reporting facilities, including trend analysis (e.g. the ability to identify the number of RFCs affecting particular CIs)

- Ease of reporting of the CI inventory so as to facilitate configuration audits
- Flexible reporting tools to facilitate impact analyses
- Ability to show graphically the configuration or network maps of interconnected CIs, and to input information about new CIs via such maps
- Ability to show the hierarchy of relationships between parent CIs and child CIs
- Interfaces easily with most systems in the IT infrastructure such as Asset, Inventory, Release, Capacity and IT Financial systems

Console Management System

Automates and streamlines the console displays such that multiple consoles can be viewed simultaneously from a single viewpoint. May also consolidate and filter console messages to reduce confusion and noise when displays are viewed.

- Provides simultaneous access to multiple system consoles in the infrastructure via networking technologies
- Filters console messages to allow for less confusion and noise when viewing them
- Allows remote users to log into various consoles without being physically nearby
- Provides enough capacity in terms of serial ports or other connections to infrastructure equipment and consoles with varying features provided by embedded software
- Provides full access to and configurability of a wide array of security protocols and encryption standards to manage and control access to consoles
- Provides options to handle cluster setups and to daisy-chain consoles to otherwise unused serial ports on nodes with other primary functions
- Allows for the capability to have users implement their own features and snap-ins using an application programming interface
- Allows users to easily transfer console displays from one display console to another either locally or remotely
- Provides support for remote port buffering
- Provides dual power supply to protect availability of displays
- Interfaces with LDAP or other directory support protocols
- Provides support for virtual clustering of consoles
- Provides similar management interfaces that can be used across multiple types of hardware and system platforms

Customer Relationship Management (CRM) System

This automates storage, handling and reporting of interactions, contacts and communications with customers of IT services.

- Provide support for customer facing operations such as face to face, phone, IM, chat, email, web and combinations of all medium interactions
- Provide interfaces for self-service kiosks and web self-services
- Maintain customer contact activity
- Provide support for different contact channels such as e-mail, web interfaces, telephony, kiosk or other avenues
- Provide contact analytics to summarize and analyze customer interactions, trends and preferences
- Provide interfaces to IT Service Desk, call agent activities and call management systems
- Provide support for targeted marketing campaigns
- Provide adequate security features to protect customer privacy and sensitive information
- Provide interfaces to IT, billing, invoicing, maintenance, planning, marketing, advertising, finance, services planning and service delivery operations
- Provide support for marketing activities such as collection of service and marketing metrics, execution of marketing campaigns, call campaigns, Web communication strategies and keeping customer relationship activities on track
- Maintain information on market share, numbers and types of customers, revenue, profitability, and intellectual property concerning customer preferences.
- Provide support for customer surveys to create, administer, collect data, analyze and report on customer satisfaction levels and preferences
- Maintain databases for customer lifecycle (time series) information about each customer and prospect and their interactions with the IT organization, including request information, support information, requests, complaints, interviews and survey responses

- Track customer intelligence to support translation of customer needs and profitability projections into strategies for different segments or groups of customers
- Provide modeling capabilities for customer relationship strategies, goals and outcomes

Database Management System

Automates control activities over the organization, storage, management, access, and retrieval of data in a database.

- Utilize a standard industry accepted data query language and set of access protocols
- Provide restart/recovery/rollback features to protect the availability, reliability and integrity of data stored in the system
- Provide data dictionary features that describe detailed descriptions of tables, schemas, record layouts and fields
- Provide indexes for quick access to data utilizing relational technologies and flexibility
- Provide support for referential integrity constraints
- Provide for stored procedures and database triggers
- Provide support for access controls such as usernames, roles, and privileges
- Provide automated storage allocation and administrative features
- Provide support for database usage reporting statistics
- Interface with other performance monitoring and data collection systems in the infrastructure
- Provide interfaces and support for report generation tools
- Provide interfaces for application calls and operation
- Provide support for transaction processing including transaction checkpoint and restart capabilities
- Provide physical view of database layout and storage use
- Provide capabilities for multiple logical views of data tied to the physical view infrastructure
- Provide backup and replication controls and features
- Provide computational features such as counts, sorts, averages grouping and cross referencing of data
- Provide rule enforcement capabilities such that rules should be able to be added and removed as needed without significant data layout redesign

Documentation Management System

This provides capabilities for creating, modifying, viewing and printing of digital documents in a variety of media formats such as word processing, spreadsheet and presentation formats.

- Provide for generation of text documents, spreadsheets and graphical presentations
- Provide schema capabilities for storing and accessing documentation by category
- Provide on-line and/or web access to documentation
- Ability to provide multi user access with check-in/check-out capabilities to prevent editing conflicts
- Ability to provide flexible authorization and authentication such as read-only access and full update access
- Provide search engine capability to find documentation
- Provide support for documenting processes and procedures such as flowcharting capabilities, information mapping and outlining
- Hypertext capability to reference associated documents
- Ability to interface to standard Corporate desktop
- Provide cut-and-paste capabilities between documents and document types
- Support version control and compliance with Corporate Documentation standards
- Provide document import/export capabilities from leading documentation generation tools
- Storage and maintenance of notification distribution lists to support ITSM activities such as Incident, Problem and Service Level Management
- Provide interfaces to company e-mail applications and services
- Flexible support for both individual as well as group notifications
- Linkages to Systems Directory
- Supports corporate security policies

Event Management System

Creates IT system events, alarms and thresholds and automates the monitoring of these to minimize human intervention. It automates communication of events of significance to support and delivery staff.

- Ability to capture and log events
- Ability to modify event status
- Ability to validate event content
- Ability to correct event content errors
- Ability to generate systems events
- Ability to filter and threshold events, i.e. ability to filter events on the basis of time, number of occurrences, etc.
- Ability to initiate actions automatically
- Ability to execute pre-defined script based on date and time (for housekeeping)
- Ability, as the result of an event, to execute pre-defined scripts and test the success
- Ability to monitor managed entity (specific metrics)
- Correlation of incidents
- Ability to forward events to distributed event managers
- Ability to collect error statistics
- Ability to generate notifications in multiple formats
- Ability to maintain local log file on managed entity
- Ability to customize parameters as required
- Interface to Configuration Management Database
- Interface to Incident Management Database
- Ability to forward events to the Event Correlation Tool for analysis and action
- Real-time status display
- Ability to access event information locally and remotely
- Provide multiple interfaces such as Command Line, Batch or GUI
- Highly flexible capability for monitoring service levels and operational level targets
- Ability to configure monitoring/monitoring scripts to summarize data for reporting purposes
- Automatic alerts for service levels in jeopardy or missed service targets

- IT component downtime data capture and recording
- Capability to alert across different communication channels such as E-Mail, Pager, FAX, or PDA
- Ability to provide variety of notification messages, memos or letters using templates for common event situations
- Ability to correlate events from different sources based on rules
- Ability to provide flexible and dynamic correlation rules
- Ability to analyze all received events
- Ability to recognize and filter repetitive events
- Ability to analyze events based on arrival time stamp
- Ability to modify event data
- Ability to accept events from different sources and protocols
- Ability to filter and threshold event data
- Ability to generate events
- Interface to Configuration Management Database
- Ability to utilize other information sources such as configuration data, i.e. ability to compare correlation tool's configuration information with actual discovered configuration information
- Ability to Initiate actions based on rules
- Ability to interface with Incident Management Database
- Ability to automatically log an Incident
- Ability to interface with other tools with open APIs
- Ability to review the Change Management Forward Schedule of Changes to determine if event from a scheduled outage/change
- Ability to send notifications in multiple formats
- Provide multi user access
- Provide flexible authorization and authentication
- Provide local and remote access
- Not restricted to a single operating system
- Ability to access event information locally and remotely

HR Management System

This automates handling of human resource activities for payroll, skills management, benefits, recruiting and reporting.

- Provide interfaces and support for company payroll processing
- Provide interfaces and support for employee time reporting
- Provide interfaces and support for employee benefits administration
- Provide database with employee contact information and other relevant personnel data
- Provide support for employee recruiting activities including tracking of recruiting efforts and coordinating between recruiters and applicants
- Produce paychecks and payroll reports
- Maintain personnel records
- Provide support for development, storage and reporting of personnel performance appraisals
- Maintain IT organization charts, reporting responsibilities and job descriptions
- Provide support for tracking skills development and maintaining skills requirements

Identity Management System

Stores, manages and administers user-ids and passwords. More advanced technologies assist with self-registration, advanced authentication support such as biometrics and single sign-on automated support.

- Provide self-service and password reset/sync interfaces
- Provide access via web services
- Interface with corporate security policies
- Provide centralized control and local autonomy functions
- Provide API interface to applications
- Workflow engine for automated submission and approval of user requests
- Provisioning engine to automate the implementation of administrative requests
- Automatic synchronization of user data from different repositories like human resources databases and enterprise directories
- Utilizes a sophisticated role-based administration model for delegation of administrative privileges
- Role and rule-based delegated administration
- Intelligent approval routing
- Auditing and reporting mechanisms

Incident Management System

This automates the workflow for logging, processing, resolving, escalating and reporting on incidents in the IT infrastructure.

- Repository to store, retrieve and report on Incidents in an easily accessible format
- Capability to link to Problem Records and RFCs related to Incidents
- Links to the Configuration Management database or Configuration Management data
- Capability for storing historical incident data and other Incident related information
- Capability to store Incident Management report data and historical reports
- Ability to allow those involved with Incident Management activities to add text to Incident records
- Ability to store and maintain alerting distribution lists
- Flexible support for desired Incident classification and logging schemas
- Flexible search capabilities
- Highly flexible routing of Incidents across control staff taking resolution and recovery actions
- Flexible support for Incident workflow activities throughout the Incident lifecycle
- Support for control staff that may be located in multiple sites or co-located in an operations bridge
- Automatic escalation facilities so as to facilitate the timely handling of Incidents and service requests
- Support and coordination for escalation and approval checkpoints
- Ability to time stamp workflow actions as they occur
- Ability to automatically issue notification alerts based on Incident type and user groups
- Linkages to Distribution List Manager
- Flexibility to tie distribution lists and notification types to Incident categories

- Automatic Incident logging and alerting in the event of fault detection on mainframes, networks, servers (possibly through an interface to system management tools)
- Automatic modifications to the Incident record being registered in order to keep control
- Interfaces to Problem Management database

Interactive Voice Response (IVR) System

Automates activities associated with servicing high call volumes to minimize human intervention with customers. Automates interfaces between voice and touch tones with a computer application that directs callers on how to proceed.

- Play recorded messages including information extracted from databases and the internet
- Route calls to Call Agents based on programmed business logic
- Provide capability to transfer callers to outside extensions
- Provide capability to collect and store caller provided information before call is transferred to a Call Agent
- Provide capability to fulfill and complete caller requests without a transfer
- Provide capability to verify caller identity before transferring the call
- Intelligently route calls using complex business logic
- Interface with corporate telephone and communications systems
- Provides APIs to interface with applications
- Interfaces with Call Management System

Intrusion Detection System

This detects unwanted manipulations of computer systems occurring through the Internet. It identifies network attacks against vulnerable applications and services, unauthorized logins and access to sensitive files. It also guards against viruses, Trojan horses, and worms.

- Ability to drive detection by security policy
- Support protection against intrusions that could result in legal liability
- Support protection against confidentiality breaches
- Support protection against email imbedded viruses
- Provide real-time protection against spy ware
- Provide automatic removal of breaches and infections for easy disposal of security risks
- Provide support for cleanup of registry entries, files, and browser settings after spy ware intrusion detected
- Support protection against spam attacks
- Support protection against degradation and loss of network services through misuse and hostile attacks
- Support protection against infection and loss of data from Web-based viruses
- Support protection against corruption of data from malicious code embedded in programming tools and scripts (i.e. Java, ActiveX, etc.)
- Prevent access to inappropriate sites by internal staff
- Provide capability to set spy ware and ad ware policies on an individual application-by-application basis
- Provide timely and periodic detection updates to stay current with latest intrusion threats
- Provide intrusion scanning facility that can be scheduled or manual
- Minimize work disruptions and slowdowns caused by scanning activities

Inventory Control System

This monitors the quantity, location and status of supplies and spare parts inventories including any related shipping, receiving, picking and put away activities.

- Monitors and tracks counts of inventory supply items such as paper, toner, cleaning supplies, maintenance kits, etc.
- Identifies locations where inventory is stored and used
- Tracks and monitors use of inventory
- Supports automated tracking of reorder checkpoints and alerts or automates purchasing systems to order additional inventory
- Provides support for inventory shipping, receiving, picking and put away processes
- Provides support for bar coding of inventory items as well as RFID (Radio Frequency Identification) tags
- Provides support for real time inventory control tracking using wireless, mobile terminals to record inventory transactions at the moment they occur
- Provides support for physical inventory counts and cycle tracking
- Provides interfaces to Definitive Hardware Store (DHS) for spare parts tracking and management

IT Financial Management System

This automates IT accounting and budgeting activities. It maintains a ledger of IT costs and revenues with associated accounting codes and cost/revenue pools. It also provides financial reports related to IT services and activities.

- Provide storage and support for IT accounting and account ledgers
- Provide support for IT cost pool descriptions and cost pool allocations
- Track, manage and report on IT accounts receivables
- Track, manage and report on IT accounts payables
- Track, manage and report on IT budgets
- Provide support for storing vendor receipts such as automated receipt logging or scanning
- Generate IT accounting reports
- Provide support for cost estimating models
- Provide capabilities for graphical reporting and multi-dimensional analysis of costs and resources
- Provide support for internal and external audit activities
- Provide flexible data analysis functions and features
- Provide storage for historical IT financial reports
- Provide support for IT financial projections and what-if analysis
- Provide interfaces to IT Billing Systems to obtain charging information
- Provide storage of historical customer bills
- Provide storage and support for IT budgets and cost models
- Provide interfaces to corporate accounting infrastructure and general ledger

IT Service Continuity Planning System

This automates the building, maintenance and communication activities associated with IT Service Continuity plans. Provides a template driven approach for documenting and building those plans to ensure they are complete.

- Provide storage and support for IT Service Continuity plans as well as plan versions
- Provides pre-built templates and requirements for building an IT Service Continuity Plan
- Supports multiple versions and releases of plans
- Tracks changes to continuity plans and compares differences between multiple plans that have been developed
- Provides interfaces to migrate plan documents to other databases and documentation tools
- Provide support for storage and management of IT Service Continuity test plans, test data and test results
- Interface with corporate Business Continuity Management data repositories
- Automatic generation of information and population of the IT Service Continuity Plan
- Provide contingency audit and review questionnaires
- Provide framework and checklist for the creation of an IT Service Continuity Plan
- Provide a Business Impact Analysis questionnaire/guide
- Provide Dependency Analysis questionnaire and guide
- Flexibility to make plan updates and changes
- Allowance for multi-user capabilities, allowing cross-departmental collaboration
- Compatibility with corporate platforms and systems
- Recognition of Vital Business Functions as key means for driving the plan

Job Scheduling System

This automates the scheduling, monitoring and execution of IT batch processing jobs.

- Ability to coordinate all shifts and production work across multiple platforms from a single point of control
- Automate complex and repetitive operator tasks
- Provide capability to dynamically modify production workload schedules in response to changes in the environment
- Manage and resolve workload dependencies
- Manage and track units of processing work
- Monitor processing work for failures and execute pre-programmed restart, backup and recovery processes based on business logic
- Notify operational staff with status of operational schedules
- Provide real-time display of schedule progress
- Provide capability for manual intervention during schedule runs if needed
- Provide capability to model schedules based on estimates in run times or business events
- Provide security functions to protect data and applications
- Provide support for creation of jobs and job schedules using job control language or operational scripts
- Interfaces with Event Management System
- Graphical user interface for inputting job schedules
- Interfaces to defined workflows and/or job dependencies
- Automatic submission of executions
- Monitor the execution of jobs
- Provides priorities and/or queues to control the execution order of unrelated jobs
- Allows for execution of alternative run schedules based on given events such as failure of the primary schedule
- Integrates real-time business activities with traditional background IT processing
- Allows for schedule operations across different operating system platforms and business application environments

Labor Reporting System

This automates and tracks labor time spent on projects, service support and delivery activities. It also reports on labor activities and usage of labor resources.

- Track and report IT labor time.
- Manage time and attendance data and IT labor profiles
- Safeguards against buddy punching or invalid recording of labor
- Interface with IT Financial Management and HR Systems
- Maintain employee numbers, project and labor codes
- Efficient and fast method for time entry
- Stopwatch timer for accurate time measurement.
- Sort time entries by project or labor code
- Filter time entries by status, day, week, month, project or labor code
- Authorizes and validates time input entries for accuracy using custom built criteria
- Multiple channels for reporting labor time such as web interface, desktop or handheld device support

Media Management System

Automates the storage, management, labeling and scratch inventories for media used to support and deliver IT services.

- Manage usage and retention of IT media such as tapes, CDs, cartridges, microfiche and removable disks
- Provide media vault management capabilities
- Manage media pick lists
- Support tagging, bar coding and labeling of media
- Assign media inventory with unique identifiers
- Manage inventory of spare media such as blank tapes, CDs, etc.
- Initialize media for use
- Protect media items from being overwritten
- Interface media management processing with Job Management systems and applications with support for file open/close interfaces
- Manage media locations such as media storage areas, spare supply cabinets and offsite storage locations
- Manage lists of media volumes that can be overwritten (scratch pools)
- Maintain an online catalog of the location of files written to media and a list of what files reside on each media volume
- Support virtualization capabilities of media creation devices
- Provide interfaces and compatibility with robotic devices for media selection and retrieval
- Capability to store and manage authorized versions of media in a secured environment
- Provide software and media license management capabilities to track usage of software and compliance with software contracting and purchase agreements
- Provide support for external software and media license audits
- Provide check-in/check-out facilities for media with access notification and status
- Provide secure access to media in line with corporate security policies

- Provide interfaces with Configuration, Change and Release Management Systems
- Provide linkages with the IT Service Continuity Management process to ensure authorized software will not get lost in the event of a major business disruption
- Provide support for software and media version control and naming/numbering standards

Network Management System

Automates monitoring, operations and administrative activities for the IT networking infrastructure.

- Monitor network to quickly identify events that represent potential incidents and problems
- Track networking resources and how they are assigned to network domains, IP and MAC addresses
- Provide support for generating and maintaining network device configurations and parameters
- Provide support for automatic reboot and configuration of networking devices
- Provide remote support access to network devices and resources
- Provide support for configuring and adding new network devices on behalf of new IT services or service requests
- Provide traffic management reporting, support and automated rerouting of traffic to different network segments
- Track usage and performance of network devices and line circuits
- Support load balancing of network traffic
- Provide support for Virtual Private Network (VPN) configurations and services
- Provide support for industry recognized communication protocols and networking services
- Log network events, usage and traffic volumes
- Provide support for monitoring and management agents within the networking infrastructure

Performance Monitoring System

Monitors usage and response times for infrastructure hardware, operating software and applications used to deliver IT services.

- Gather key metrics for the hardware components (mainframe, network, server, and workstation):—utilization, throughput, workload, buffers, response
- Gather key metrics for the software components (system, middleware, application): utilization; throughput; buffers; queues; response
- Consolidate metrics by time, group of components
- Ability to interface to standard corporate desktop
- Use standard event protocol to communicate event status
- Ability to interface with System Event Monitor
- Capability to run scripts including unattended
- Capability to send alarms/alerts
- Ability to measure/determine idle cycles
- Ability to capture data by machine
- Ability to capture data by process
- Ability to correlate cause and effect analysis
- Ability to work in exception more
- Access to application configuration info
- Access to historical component database
- Centralized, real-time, and historical trending monitoring capability
- Collect and consolidate data into central repository—organize by CI
- Consolidate capacity data from individual components
- Supports start and stop criteria
- Trace hooks (API) for applications
- Web interface
- Ability to record task ownership and restrict access by owner
- Common front end for parameter setting

Portfolio Management System

This automates the management of IT strategy initiatives through systematic tracking of project and service initiatives considering their costs, revenue current and strategic impacts.

- Define project and product portfolio types for use in planning projects and services
- Providing metrics for evaluating services and projects
- Report on services and projects for net present value, costs, benefits, return-on-investment and risk
- Provide capability to customize project and services portfolio inventory and analysis criteria by adding new fields, creating new pages, creating new templates, rearranging forms and adding custom formulas
- Provide capabilities for portfolios to be grouped in hierarchies that reflect the structure of the business or decision-making process
- Connect service and project portfolio decisions to investments and service lifecycle process such as defining decision gates for go-ahead decisions
- Provide capabilities for what-if scenarios, scenario comparison and scenario modeling
- Provide service and project dashboards
- Provide capabilities to customize dashboards and reports
- Provide drill-down capabilities for investment details
- Provide highlighting capabilities to quickly identify problem projects, and troublesome service investments based on customized criteria
- Integrate portfolio elements to IT Financial and Project Management Systems
- Integrate portfolio elements with IT Service Catalog

Problem Management System

This automates the workflow for logging, processing, resolving, escalating and reporting on problems and Known Errors in the IT infrastructure.

- Repository to store, retrieve and report on Problems and Known Errors as well in an easily accessible format
- Repositories for the collection of appropriate incident/ incident lifecycle data and information
- Capability to automatically generate and link to RFCs related to error removal
- Links to the Configuration Management database or Configuration Management data
- Capability for storing historical incident data and other problem related information
- Capability to store problem management report data and historical reports
- Ability to allow those involved with Problem Management activities to add text to Problem records
- Links to Incident Management database to obtain incident data to assist with problem identification and resolution
- Links to Incident Management database to communicate information on Known Errors to Incident staff
- Ability to link Incident records with Problem records
- Storage and management of Known Error records
- Storage and retrieval of resolution and recovery actions associated with Known Errors to avoid repeating those actions again for similar Incidents
- Highly flexible search capabilities
- Linkages to Incident, Configuration and other repositories that can provide further actions or detail as needed to research Problems

Process Modeling Tool

This automates the creation, maintenance and simulation of processes to support process design and execution activities.

- Build and define business and process models
- Represents both the current process state, future process state and gaps in between them
- Graphical modeling interfaces to quickly develop process models using a variety of shapes and graphics
- Model inputs and outputs of processes
- Model dependencies between processes
- Allow for customized process labels and descriptions
- Utilize a variety of process description methods such as flowcharting, activity diagrams, LoveM charting, networking and IDEF standards
- Simulate outcomes from process execution
- Identify process bottlenecks and queue lengths
- Model process timing outcomes from process execution
- Provide interfaces and links to technologies employed with the process that can be used to guide those using the process with those technologies
- Generate process documentation in a variety of formats from models that are generated
- Provide nesting and hierarchical modeling capabilities
- Supports industry standard modeling languages such as Business Process Modeling Notation (BPMN), Business Process Execution Language (BPEL), Unified Modeling Language (UML), Object Process Methodology (OPM), and Web Services Choreography Description Language (WS-CDL)

Procurement System

This automates the tasks, workflows and storage of information related to the purchase of IT service assets.

- Generate purchase orders in alignment with business procurement standards
- Maintain supplier contact information
- Store and maintain versions of generated proposals, contracts, RFPs, RFIs and vendor cost estimates
- Record and maintain vendor service contracts
- Support procurement processes with workflow and approval gates
- Report on status of procurement requests and purchase orders
- Generate work orders for approved service purchases
- Support vendor selection activities using customized vendor and product criteria
- Track change orders related to purchases
- Track receipt of goods and services
- Record add-on charges incurred during receipt of good and services
- Track purchase receipts
- Copy existing purchase orders to make new ones
- Maintain database with benchmarked costs for common purchases
- Interface with IT Financial Management, Billing and Inventory Control Systems
- Support E-procurement activities

Project Management System

Automates tasks associated with managing projects such as project scheduling, cost control, budget management, resource allocation, status reporting, documentation and project administration.

- Provides for entry and storage or task information such as task names, descriptions, start dates, end dates and resource allocations
- Provides facility for inputting mass data such as importing project information from spreadsheets
- Web support features to access project information over the web
- Collaborative project support features to support multiple team members working with project management tasks
- Support for collaborative building of project plans
- Clearly structured user interface
- Support for identifying and managing dependencies between project tasks
- Support for key project management practices such as those described in Prince2 or PMI such as Earned Value Analysis reporting
- Analysis of project critical path and critical path dependencies using a user selected variety of standard techniques such as PERT, CPM and MPM
- Support for versioning of project plans
- Ties project milestones and deliverables to project tasks
- Support project resource management such as automatic assignment of resources to tasks, part time resources, indication of over-assignment of resources and ability to document detailed listing of assigned tasks by resource
- Support for identifying and reporting on Work Breakdown Structures
- Provides GANTT Charting
- Support for modeling of impacts to changes in project tasks, schedules and resources

- Program Management support
- Support for project costing and estimating as well as providing interfaces to the IT Financial Management System

Prototyping System

This provides rapid prototyping of application solutions to capture solution flow, customer acceptance and service design requirements. It generates code elements from design prototypes.

- Rapidly draw applications and screens with use of widgets such as menu bars, hyperlinks, scroll bars, files and databases
- Rapidly create application screens and logic flows
- Simulate application workflows and operation
- Support full screen mode for application simulations
- Allow for dynamic changes to application flows and work elements for which their impacts can be quickly simulated
- Flow created simulations and prototypes into requirements documents
- Create custom widgets
- Export user interface specifications to standard Documentation Management tools
- Undo and redo changes to screens, flows and descriptions
- Provide robust backup and recovery for generated prototypes and simulations
- Generate code elements from developed prototypes
- Manage and maintain multiple versions and iterations

Reader Board System

Provides graphical or textual display of IT infrastructure events and messages in a manner where they can be seen by all support staff in processing and command centers.

- Provide display capability that allows messages to be viewed by technology center staff
- Allow for manual entry of messages
- Interface with System Event Monitors and call management infrastructure
- Provide enhanced display to notify technology center staff based on message severity or impact
- Audible alarm capability for critical incidents
- Provide local centralized messaging capabilities to signs and boards that may be in remote locations

Release Build Manager

This automates the workflow and activities used to bring designed services into the production operations environment. It supports building, testing, naming and packaging of application and service releases.

- Supports automated build of new Releases of software applications for mass distribution and use
- Ability to drive program compilations and links, in the correct sequence, under program control using the correct versions of the source code as stored in the DSL
- Ability to generate master CD-ROMs or other media version of software that will be replicated for mass distribution
- Ability to support use of the cross-reference information stored in the Software Configuration Management tool to determine which parent CIs need to be rebuilt when lower-level units are changed
- Repository to store, retrieve and report on Releases in an easily accessible format
- Capability to link to Problem Records and RFCs related to Releases
- Links to the Configuration Management database or Configuration Management data
- Capability to store Release Management report data and historical reports
- Ability to store and maintain distribution lists for staff related to release activities
- Flexible search capabilities

Remote Support Tool

Provides automated capabilities to operate, configure and support IT service assets at remote locations without requiring physical human presence at those remote locations.

- Remote diagnostic capability to gather information related to Incidents at other sites
- Ability to monitor user sites as necessary to gain views into symptoms
- Ability to assume control over user terminals, workstations and server operations to assist in Incident resolution and recovery actions
- Real time chat facility
- Secure file transfer capability
- Ability to view remote terminal screens and consoles
- Shared web browsing capability
- Remote-to-local printing without the installation of additional drivers
- Ability to remotely edit registry, schedule reboots, manage users, and administer processes and services
- Ability to see remote CPU/file usage and virtual memory settings
- Support for command line installation and scripted mass-deployment
- Support for two direction file transfer and automatic folder synchronization
- Ability to provide support access and control from handheld devices
- Ability to operate within corporate security policies
- Auto-resolution of dynamic IP addresses for seamless connectivity
- Ability to provide remote support without needing to reconfigure firewalls or other secure perimeter infrastructure
- Remote control of operations for servers and desktops, for example, to assist with making Changes to a server as part of a Release rollout
- Remote monitoring of the event logs and other Problem logs on servers

- Remote monitoring of processor, memory and disk utilization
- Remote management of the disk space on servers—for example to monitor usage, to reorganize files for improved performance, and to allocate more disk space to releases
- Ability to restrict changes that individual Users can make to client workstations to make the target destination for new Releases much more reliable

Report Generator

This automates the generation of reports with a wide variety data extraction, analytical and manipulation methods.

- Data search capabilities
- Flexible data extraction capabilities
- Report generation capability
- Statistical analysis on stored data
- Automatic generation of management and trend information relating to Incidents
- Ad-Hoc reporting capabilities
- Flexible data formatting and extraction capability
- Ability accept different input formats
- Ability to provide attachments to reports
- Ability to import/export data using open APIs
- Ability to utilize corporate documentation standards
- Web reporting capability to allow distribution of reports on the corporate Intranet
- Stores and maintains report distribution listings

Request Management System

This automates the workflow for logging, processing, fulfilling, escalating and reporting on IT service requests.

- Workflow model for managing work requests to fulfillment
- Customized forms for entering requests
- Email notification of request completion to request owners and fulfillers
- Online request order catalog facilities
- Timed reminders and auto generation of repeated requests
- Request reporting and status dashboard showing status of requests and fulfillment activities
- Interfaces to Change Management System
- Online presentation interface that allows remote entry of requests
- Ability to track each request from open to close.
- Capability for requestors to view status of request tickets
- Responses to requests can be sent to multiple recipients simultaneously
- Generation of detailed and customized reports for IT management
- Interfaces to Procurement and IT Financial Systems
- Capabilities to limit requests only to authorized requestors
- Capabilities for requestors to view only requests authorized to make
- Maintain catalog of standard requests
- Provide support Frequently Asked Questions capability
- Maintain and report on request ticket history
- Request report and full search capabilities
- Self-Service web interface
- Email submission interface
- Email interface for ticket updates
- Support for categorizing and prioritizing requests
- Time stamping for request tickets to mark fulfillment milestones and activities
- Support for manual and automated approvals

Security Management System

Provides automated capabilities to design, implement, administer, monitor and maintain the IT security infrastructure to support audit and security policy objectives.

- Raise and escalate security related events to IT staff based on security policy and business logic
- Support for a wide variety of system and networking platforms, devices and file systems
- Detection and logging of file additions, deletions and changes to content, permission and attributes
- Detection of changes to file directories and directory permissions and attributes
- Provide rule base capability for security event detection and escalation
- Automatic archiving capability to store changes in the infrastructure to minimize record keeping and satisfy audit objectives
- Import functionality for external asset and inventory lists to minimize manual entry
- Identification of device configurations highlighting missing configuration parameters
- Enforce separation of duty to ensure that only authorized personnel can designate baselines of the infrastructure
- Interface to Incident Management Database
- Automate scans of the infrastructure to reduce the cost and time associated with manual security checks
- Maintain, implement and track security patches and fixes
- Safeguard against host and network intrusions and viruses
- Log security incidents
- Provide support for forensics activities when analyzing security incidents
- Provide reports that identify security incidents and breaches
- Maintain, implement and operate firewall rules

Security Test Manager

This automates activities to test for security vulnerabilities in the IT infrastructure.

- Network Scanning capability
- Vulnerability Scanning capability
- Password Cracking capability
- Log Review analysis and analytics
- Integrity Checker capabilities and features
- Virus Detection testing capability
- War Dialing
- War Driving (802.11 or wireless LAN testing)
- Infrastructure Penetration Testing capability
- Forensics support and analysis
- Investigate and disconnect unauthorized hosts when found
- Harden infrastructure by disabling vulnerable systems and services
- Test compliance with established security policies
- Validate timeliness and current levels of security patches and fixes
- Limit or block access to unwarranted IP addresses when found
- Validate virus definitions are current and up to date
- Test for wireless vulnerabilities
- Interface with Event and Incident Management Systems
- Validate vulnerability to denial-of-service attacks
- Flexibility to limit penetration tests to specific IP addresses or ranges
- Logs and reports on test results

Service Catalog Manager

This automates the creation, storage, communication and maintenance of service descriptions and related information for IT services delivered to customers.

- Stores service descriptions and related information
- Links related services into service lines or bundles
- Manages and reports on status of services (i.e. in production, in transition, proposed, retired, etc.)
- Web enabled catalog interface
- Security features to restrict access or services viewed to only those authorized to see them
- Identifies and maintains links and dependencies between services
- Interfaces with Request, Change, Configuration and Project Management Systems
- Interfaces with IT Financial and Billing Systems
- Highlights changes to service descriptions and features
- Maintains historical archive for service descriptions that have been retired
- Attractive user interface for finding and selecting services
- Search capabilities to find services by keyword

Service Knowledge Management System

Supports the identification, creation, representation and distribution of knowledge used to support and deliver IT services. Provides centralized access to this knowledge such that information can be searched for and obtained quickly. Also manages knowledge in a variety of formats.

- Store, track and maintain knowledge documents and artifacts
- Presentation graphical front-end to provide easy access to knowledge documents and artifacts
- Keyword search capabilities to find documents or lists of documents
- Interfaces with Documentation Management System
- Role based security interfaces to ensure knowledge is seen only by those authorized to see or maintain it
- Check-in/Check-out controls to make sure updates to knowledge artifacts do not cause conflicts
- Support and control over artifact naming conventions
- Support and control over artifact versioning
- Support for hierarchical document structures to structure knowledge with ability to move documents between hierarchies
- Quick link ability to guide users to other portals and web sites
- Storage capacity management functions to ensure enough storage and bandwidth capacity exists for storing and accessing knowledge
- Built-in administrative features that allow for administrative tasks such as managing access to knowledge, activating and deactivating users, maintaining artifacts and security access controls
- Interfaces with a large variety of databases and document types
- Operational reporting features to report on knowledge access, availability and usefulness

Service Level Management System

This automates the workflow for logging service agreements, managing them and reporting on IT service levels and service quality. Also reports and maintains status on service improvement projects, plans and activities.

- Store SLAs, OLAs and Underpinning contracts
- Capability to support a variety of SLA structures such as master SLA agreements with extensions for unique business unit needs
- Capability to store service data such as SLA/OLA results
- Capability to maintain service historical data and information
- Interfaces to Customer Relationship Management (CRM) System
- Ability to tie SLA/OLA agreements with business units and departments
- Ability to store service communication distribution lists for service reporting
- Provides linkages between Service Catalog Manager and SLA/OLAs
- Interfaces with Report Generator to produce service reports
- Provide linkage to Procurement System with underpinning contracts
- Reports or provides on-line display of services and their quality status
- Provides an overall dashboard or scorecard of IT service quality
- Indication as to whether service levels are being met
- Highlights service issues and failures
- Provides different management, service and operational views
- Provides drill-down to sub-services and other information sources
- Linkages to Service Level Management Database
- Interfaces to Event Management System
- Provides analytics to review and manage quality of services delivered

- Ability to generate service workflows by SLA/OLA and Service Catalog categories
- Support for workflows to build, agree, approve and maintain SLA/OLAs
- Linkages to Service Level Management Database
- Support for multiple generations of service agreements
- Flexible support for service agreement structures such as a base service agreement with unique business unit addendums
- Logging of negotiation and contact status
- Integrated to-do lists with calendaring
- Flexible maintenance support for SLAs, OLAs and UCs
- Integration with Service Catalog Manager

Software Configuration Manager

This automates the workflow, tasks and packaging associated with assembling release components into release units.

- Manages the different versions of software source code during its development
- Manages relationships between software components to identify software CI changes for impact upon other parts of the release
- Flexible and configurable support for release packaging requirements such as delta, full and packaged release units
- Linkage with Change Management and Problem Management systems to link software CIs and release units back to a change or problem
- Supports automatic assembly of release components for compile, build, link-edit to executables
- Provides notification of incomplete software assemblies
- Interfaces with Release Build Manager
- Interfaces with Configuration Management System
- Controls releases and their changes
- Records and reports on the status of release components
- Ensures completeness and consistency among release components.
- Manages the process and tools used for release builds
- Supports adherence to the organization's development process
- Provides software defect tracking and makes sure defects have traceability back to the source

Software Distribution Manager

This automates the workflow, tasks and activities associated with distributing software release packages to the IT infrastructure in a controlled manner that reduces the need for human intervention.

- Audits and assures delivery of software files
- Integrity checking of data sent and the ability to restart broken transmissions from the point of failure
- Variety of delivery options to optimize the usage of network capacities such as push, pull, staging and promotion schemes
- Courier mechanism via CD-ROM, for sending whole packages to remote installations not part of a network
- Fan-out capabilities to employs intermediate servers at remote locations to help with distribution activities
- Ability to store a new version of an application in a dormant state to support rapid activation or deactivation when triggered
- Interfaces with Release Build Manager

Staffing Calculator

Provides automated capabilities to model, estimate and forecast staffing head counts needed to support and deliver IT services.

- Capability to estimate number of staff required to provide IT services for various functions such as Service Desk, Technical and Operations Management
- Base results on forecasted volumes of events, tasks, estimated task durations and productive employee time per day
- Provide indication of staffing shortages
- Estimate delays based on input volumes and number of staff
- Provide what-if scenarios for modeling staffing plans and key events that may impact planned labor volumes
- Provide capability for storage of multiple modeling scenarios

Storage Management System

This automates the management and operation of the IT storage infrastructure including creation, deletion, access, archival and modification of physical storage assets within the IT infrastructure.

- Manage, retain and report on digital data within time limits that meet business needs for data retention and compliance
- Manage files and file structures from a logical perspective without manually maintaining storage locations, allocations and addresses
- Provide access to digital data through both sequential and random means
- Provide non-disruptive migration of data volumes within the storage subsystem
- Provide support for externally attached storage systems
- Provide support for direct attached, network'attached or storage area networks (SANs)
- Provide reporting capabilities for storage availability and use
- Interface with Event Management System
- Provide search capabilities to define tiers within the virtualized storage pool managed by the storage system
- Provide capabilities to identify, classify. and move volumes to match price, performance and availability application requirements
- Defines groups of logical storage volumes that can be migrated together between tiers
- Integrated with Performance Monitoring and Capacity Modeling Systems to use information in planning data migration requirements
- Command line and scripting interface to handle general storage tasks such as volume provisioning
- Provides support for data retention standards and compliance issues
- Support safeguards against unwarranted storage use or access

- Supports high performance access such as caching or flash memory
- Provides interfaces to automated backup and efficient restore of data files
- Provides compatibility with existing infrastructure IO subsystems, channels and operating systems
- Provide interface and compatibility with Job Control System

Surveillance System

This provides facilities to monitor the behavior of staff, service assets or physical facilities for security and access control.

- Provides facilities such as Closed Circuit Video Television monitoring (CCTV) to observe staff areas, data centers, internal and external physical locations from a centralized control facility
- Provides real-time capture of surveillance data
- Provides quick search of captured surveillance media to find specific events by time, motion event or date
- Allows a variety of operational modes such as continuous monitoring or only as required to monitor a particular event
- Provides for long-term storage and archiving of recordings on media
- Allows surveillance media to be re-usable such as tape that may be cycled through the recording process at defined intervals
- Provides flexibility in quality of media recordings by custom adjustments to data compression ratios, images stored per second, image size and duration of image retentions
- Provides for either analogue or digital means for surveillance data capture
- Provides centralized display monitors for viewing surveillance capture
- Interfaces to Building Management System, PC desktop or server to handle storage of surveillance media and raise surveillance events
- Password and authorization protection to protect unauthorized access to surveillance systems and media
- Provides for both audio as well as video surveillance
- Centralized management software features to view and manage multiple sources of surveillance capture devices from a single location
- Provides for indoor as well as outdoor surveillance
- Provides weather proofing for outdoor surveillance devices

- Interfaces to audible alarm system for a variety of captured surveillance events
- Provides wide variety of surveillance devices for unique location needs such as imbedded cameras that are not easily visible in surveyed areas
- Compatible with industry networking standards and protocols for data transfer activities such as video streaming

System Directory

Stores and organizes information about a computer network's users and network resources, and that allows network administrators to manage users' access to the resources. This also acts as an abstraction layer between users and shared resources.

- Provide a single point of management for all infrastructure user accounts, devices and applications
- Provides support for X.500 protocols and distributed directory services models
- Provides support and capabilities for namespaces
- Authenticates system resources that manage directory data
- Provide an authorized place to store information about network-based entities, such as applications, files, printers, and people
- Manage identities and broker relationships between distributed resources
- Interface with management and security mechanisms
- Provides consistent standard for naming, describing locating, managing, accessing and securing individual resources
- Provide interfaces and synchronization support for other system directories used by the company
- Organize information hierarchically to ease network use and management
- Flexibility to store a wide range of attributes in the directory and tightly control access to them at the attribute level (i.e. allow global access to a person object, but lock access to the Social Security Number attribute)
- Allow for creation of multiple copies of the directory and automatically replicate and synchronize changes among them
- Support multi-master replication for flexibility, high-availability, and performance
- Provide role-based delegation of administration rights to let administrators delegate specific administrative privileges and tasks to individual users and groups

- Provide highly flexible directory search and query functions
- Provide Internet-ready security services to protect data while facilitating access
- Support fully integrated public key infrastructure and Internet secure protocols (i.e. LDAP over SSL) to let organizations securely extend selected directory information beyond firewall perimeters
- Provide open synchronization mechanisms to ensure interoperability, replication and synchronization with multiple platforms

Test Data Generator

This automates generation of test data for testing applications and services using customized criteria.

- Ability to generate data records and populate data fields based on complex business logic
- Provide command line and GUI interfaces
- Flexible data record formatting functions
- Populates data fields based on complex business logic
- Storage for multiple versions of test data
- Provision of listings of test data
- Ability to create volumes of unique records automatically without re-entry
- Capability to audit data generated to ensure unique key fields have been populated without replicated data
- Ability to automate the creation of test data to match test criteria
- Ability to simulate processing volumes and loads for stress testing
- Ability to manage the creation of test criteria and log results against criteria to support test reporting and status
- Ability to simulate end-user keystrokes and/or other actions that would simulate use of the live system

Test Management System

Automates many of the activities involved with testing IT services and applications such as test planning, test case creation, test execution, bug tracking and test results reporting.

- Supports development of test plans
- Supports scheduling of testing activities
- Assigns and manages testing personnel
- Interfaces with Test Data Generator to build and maintain test data
- Tracks and reports on test execution results
- Provides for defect reporting
- Supports features that keep all test data in a centralized repository with real time data that facilitates instant collaboration
- Maintains and updates Known Errors and bug lists for releases that under test
- Allows for creation of test scenarios with scripted test runs and expected results
- Compares expected results to actual results achieved during testing
- Maintains all test data in a centralized repository with real time data that facilitates instant collaboration
- Intranet/internet support for test activities that take place in local and remote locations
- Support for parallel execution of test cases by test team personnel that are globally spread apart
- Tracks individual testing accomplishments over any range of time such as projects and releases worked on, number of test cases written and executed, and filed defects
- Provides automated dashboards for key test results and testing progress
- Fosters a structured test process with relevant status reporting to facilitate planning, test execution, results tracking, and release decisions throughout the software development lifecycle

Uninterruptible Power Supply (UPS) System

This provides a steady source of clean uninterruptible electrical power with failover to a secondary power source if the primary source is interrupted.

- Provide uninterruptible power supply at levels that meet requirements for availability and recovery
- Provide primary and secondary power sources with the ability to switch to the secondary source if the primary fails
- Capability to switchback to the primary power source from the secondary when the primary source becomes available
- Capability to operate using the secondary primary source for a length of time needed to maintain or recover service operations
- Selected type matches availability and recovery needs (i.e. Battery Backup, Line Interactive, Ferro Resonant Standby, etc.)
- Protects against power spikes and drops with surge suppression, filtering and line conditioning functions
- Converts AC power to DC power and DC Power back to AC power
- Form factors meet requirements for data center floor space and HVAC such as standalone or rack mounting
- Sizing factors meet capacity and run time requirements to support availability and recovery
- Provide capabilities for adding expandable units to increase capacity and run time at later points in time to meet changing power and equipment needs
- Provide self-test capabilities with interfaces to the Incident Management and Building Management Systems
- Automatic shutdown capabilities that make use of control hardware and software that interfaces to infrastructure operating systems
- Operating range that provides uninterruptible service in cold and warm extremes
- Battery should be user replaceable

- Equipment should be industry certified for use and operation
- Equipment should contain visible LED displays to quickly check for normal operations

Chapter

4

Data Architecture

Data Architecture Overview

This chapter provides a data view of the IT Service Management architecture presented in the previous chapter. The data view presented here consists of data building blocks described as generic database types.

Each generic database type is listed with a high level description and a set of configuration items that would be associated with it. These are the same CIs described in the functional architecture. They are specified here again to help describe more detail about the kinds of information associated with each generic database type.

A key use of the data architecture here is to represent all needed data items and repositories that should be inherent in the IT Service Management architecture. These can also form the basis for a comprehensive Service Knowledge Management System (SKMS) by adding a portal and presentation interface with security features to access all the information used to run the infrastructure.

Access Management Database

Holds data related to user-IDs, passwords and permissions to service assets in the IT infrastructure.

Logical CIs represented in this database include:

- System Directories
- User-IDs
- Passwords
- Security Roles and Permissions

Architecture Database

Holds data related to infrastructure technology standards and technology frameworks.

Logical CIs represented in this database include:

- Architecture Standards
- Data Architecture
- Documentation Templates
- Enterprise Architecture
- IT Research
- Management Tooling Architecture
- Naming Standards
- Network Architecture
- Organization Architecture
- Process Architecture
- Service Compliance Requirements
- Technical Architecture
- Tooling Requirements

Asset Management Database

Holds data related to service asset inventories and their financial aspects.

Logical CIs represented in this database include:

- Asset Configurations
- Asset Inventory Reports
- Asset Listings
- Asset Records
- Asset Tags
- Definitive Hardware Store Inventory (DHS)
- Equipment Warranty Records
- Packing Slips
- Supply and Inventory Control Policies
- Systems Software Inventories

Availability Management Database

Holds data related to risks, plans, requirements, and availability configurations to ensure that IT services will be available when needed.

Logical CIs represented in this database include:

- Availability Database
- Availability Plans
- Availability Report
- Hardware Maintenance Requirements
- Risk Assessment
- Risk Register
- Software Maintenance Requirements
- Vital Business Function (VBF) Inventory
- Vital Business Functions

Call Management Database

Holds data related to Service Desk calls, calling status and calling flows.

Logical CIs represented in this database include:

- Automatic Call Distribution Systems (ACD)
- Call Handling Report
- Call Management Software
- Service Desk Call Handling Statistics
- Service Desk Software
- Toll-Free Access Lines

Capacity Management Database

Holds data related to business demands, workload characterizations, equipment usage, equipment forecasts and plans needed to ensure enough capacity is in place to handle service demands.

Logical CIs represented in this database include:

- Application and Service Sizing Estimates
- Business Forecasts
- Capacity Baselines
- Capacity Data Analytics
- Capacity Database
- Capacity Models
- Capacity Plan
- Capacity Report
- Capacity Thresholds
- Demand Factor Descriptions
- Demand Forecasts
- Demand Strategies
- Disk Utilization Reports
- Network Traffic Reports
- Resource Forecasts
- Server Utilization Reports
- Storage Usage Reports
- System Performance and Utilization Data
- System Performance and Utilization Reports
- Workload Characterizations

Change Management Database

Holds data related to changes in the infrastructure and their status.

Logical CIs represented in this database include:

- CAB Meeting Minutes
- CAB/ECAB Member Listings/Distribution Lists
- Change Approval Packages
- Change Historical Reports
- Change Status Reports
- Change Tickets
- Forward Schedule of Changes (FSC)
- Planned Service Availability (PSA) Notices
- Post Implementation Review (PIR) Findings
- Requests For Change (RFCs)

Configuration Management Database

Holds data related to service assets, their configurations and relationships.

Logical CIs represented in this database include:

- Access Circuits (Lines)
- Anti-Virus Subscription Lists
- Application Code Library Configurations
- Biometric Devices
- Cabinets and Racks
- CD Drives
- Channel Subsystems
- Circuit Numbers
- Collators and Binding Equipment
- Command Center Configurations
- Configuration Audit Results
- Configuration Databases
- Configuration Item Naming Conventions
- Configuration Item Records
- Configuration Reports
- Consoles
- Controllers
- Copy Machines
- Data Dictionaries
- Data Record Layouts
- Databases
- Definitive Media Library (DML) Inventory
- Definitive Software Libraries
- Desktops
- Device Configuration Images
- Disk/RAID Configurations
- Domain Naming Services (DNS)
- Electronic Memory Devices
- Equipment IP Addresses
- Equipment Physical Configuration Specifications
- Escalation Contact Lists
- Extranet Configurations
- Fax Devices

- Federated Security Configurations
- Files
- Firewall Appliance Devices
- Firewall Configurations
- Firewall Servers
- Firewalls
- Handheld Devices
- Hardware Virtualization Configurations
- Hubs
- Include and Assembly Configurations
- Intranet Configurations
- IP Addresses
- IP Addressing Schemes and Configurations
- IP Telephony System
- IP/MAC Addresses and Owners
- Laptops
- Local Area Network (LAN) Configurations
- Logical Data Models
- Mainframes
- Management Information Base (MIB)
- Microwave Configurations
- Modems
- Monitors
- Network Adapter Cards
- Network Appliance Devices
- Network Device Configurations
- Network Topology Maps
- Patch Panels
- Point Of Sale Devices
- Policy Enforcement Systems
- Port Assignments
- Portable Disk Drives
- Printers
- Privacy Configurations
- Proxy Configurations
- Proxy Configurations
- Proxy Servers
- Public Branch Exchange (PBX)
- Reader Displays
- Remote Support Systems
- RIM Devices

- Routers
- Satellite Configurations
- Scanners
- Security Scanning Devices
- Sensors
- Servers
- Service Models
- Software Configuration Control Systems
- Software License Keys
- Software Licenses
- Storage Adapter Cards
- Storage Area Networks (SANs)
- Storage Compression Schemes
- Storage Configuration Layouts
- Storage Devices
- Storage Management Systems
- Switches
- System Consoles
- Telephone Numbers
- Telephones
- Test Lab Layouts and Configurations
- Transaction Control System Configurations
- Video Conference Equipment
- Video Surveillance Devices
- Virtual Private Network (VPN) Configurations
- Wide Area Network (WAN) Configurations
- Wireless Access Equipment

Customer Relationship Management (CRM) Database

Holds data related to contacts with customers that use IT services, their requirements and needs.

Logical CIs represented in this database include:

- Customer Contact History
- Customer Sales Calls
- Customer Satisfaction Surveys
- Customer Satisfaction Survey Results
- Market Research
- Relationship Contact Listings
- Relationship Issues and Status
- Service Brochures
- Service Dashboard
- Service Marketing Plans
- Service Metrics
- Service Quality Reports
- Service Quality Survey Results
- Service Quality Surveys
- Service Reports
- Service Review Meeting Notes

Definitive Media Library

Holds definitive authorized versions of media with configurations and software used to build, transition, and operate the IT infrastructure.

Logical CIs represented in this database include:

- Application Code
- Application Code Generators
- Application Development Software
- Application Programming Interfaces (APIs)
- Asset Management Software
- Backup and Restore Tools
- Capacity Data Collection Applications and Scripts
- Change Management Software
- Chargeback Software
- Checkpoint Restart and Control Software
- Code Compilers
- Configuration Management Software
- Database Management Systems (DBMS)
- Distribution List Management Software
- Documentation Support Software
- Event Monitoring Software
- File Transfer Software
- Hardware Maintenance Mgt Software
- Incident Management Software
- Interactive Voice Response Applications
- Intrusion Testing Software
- IO Subsystems (IOS or BIOS)
- Media Archive Control Software
- Media Management Software
- Middleware/Messaging Software
- Network Control Systems
- Operating Systems Software
- Patch Management Support Software
- Performance Monitoring Software
- Physical Plant Equipment Documentation
- Problem Management Software
- Process Modeling Tools

- Security Application Protocols
- Security Applications
- Software Maintenance Mgt Software
- Specialized Device Control Systems
- System Utilities
- Tape Management Software
- Transaction Control Software
- Virtualization Software
- Vulnerability Testing Software

Event Management Database

Holds data related to events in the IT infrastructure and how they get processed.

Logical CIs represented in this database include:

- Auto Discovery Systems
- Building Management System Event Logs
- Event Correlation Schemes
- Event Definitions
- Event Descriptions
- Event Filtering Criteria Lists
- Event Handling/Escalation Lists
- Event Logs
- Event Status Reports
- Event/Alarm Thresholds
- Monitoring Agents

Facility Management Database

Holds data related to physical facilities used to house and protect service assets in the IT infrastructure.

Logical CIs represented in this database include:

- Building Management Systems
- Command Center Layout Diagrams
- Cooling Chillers
- Facility Blueprints
- Facility Building Code Requirements
- Facility Cooling Diagrams
- Facility Electrical Safety Codes
- Facility Fire Safety Codes
- Facility Floor Plan Layouts
- Facility Physical Requirements
- Facility Wiring Diagrams
- Fire Suppression Systems
- Mechanical/Electrical Layouts
- Office Floor Layout Plans
- Premise Cabling Diagrams
- Uninterrupted Power Supply (UPS)

Financial Management Database

Holds data related to IT financial budgets, costs and charges.

Logical CIs represented in this database include:

- Account Code Lists and Descriptions
- Budget and Expenditure Forecasts
- Budget Line Item Descriptions
- Budget Reports
- Charging Algorithms and Calculations
- Charging Code Lists and Descriptions
- Charging Summary Report
- Cost Estimates
- Cost Models
- Cost Pool Descriptions
- Purchase Orders
- Service Catalog
- Service Charging Invoices
- Staffing Estimates and Projections
- Time Reports
- Vendor Invoices

HR Management Database

Holds data related to human resource assets, policies and role/job descriptions for IT personnel.

Logical CIs represented in this database include:

- HR Policies
- HR Recruiting Logs
- Job Descriptions
- Organization Chart
- Personnel Records
- Role Descriptions
- Roles and Responsibilities Matrix
- Skills Inventory

Incident Management Database

Holds data related to incident records and the status of incidents that have occurred in the IT infrastructure.

Logical CIs represented in this database include:

- Incident Classification Schema
- Incident Notification Lists
- Incident Status Reports
- Incident Tickets
- Repair Dispatch Tickets

IT Service Continuity Management Database

Holds data related to IT Service Continuity plans, schedules and test results.

Logical CIs represented in this database include:

- Business Continuity Plans
- IT Service Continuity Plan
- IT Service Continuity Planning Systems
- IT Service Continuity Test Plan
- IT Service Continuity Test Results
- IT Service Continuity Testing Schedules

Knowledge Management Database

Holds data related to knowledge used by staff to support their IT service support and delivery activities.

Logical CIs represented in this database include:

- Activities
- Business and IT Notices and Announcements
- Business Presentations
- Documentation Subscription Lists
- Facility Emergency Exit Procedures
- Facility Locations
- Hardware and Software Benchmarks
- Holiday Schedules
- How-To Help Knowledge
- IT Audit Results
- Management and Line Staff Presentations
- Meeting Agendas
- Meeting Minutes
- Physical Equipment Maintenance Manuals
- Procedures
- Process Descriptions
- Process Guides

- Process Metrics
- Process Models
- Research Reports
- Research Subscriptions
- Security Alert Notices
- Training Guides
- Vendor Manuals
- Whitepapers and Articles of Interest

Known Error Database

Holds data related to Known Errors in the IT infrastructure.

Logical CIs represented in this database include:

- Known Errors

Operations Management Database

Holds data related to operational activities that underpin the day-to-day operations of services in the IT infrastructure.

Logical CIs represented in this database include:

- Backup Schedules
- Customer Contact Lists
- File Transfer Schedules
- File Transfer Verification Logs
- HW/SW/NW Maintenance Procedures
- Job Control Logs
- Job Run Schedules
- Job Schedule
- Job/Script Control Data (JCL)
- Maintenance Records
- Offsite Storage Archival Pick Lists
- Operational Plan
- Operational Run Books
- Operational Run Books
- Operational Run Logs
- Operational Scripts
- Operational Scripts
- Operations Staffing Plan
- Process Owner Contact List
- Process Quality Reports
- Repair Service Schedules
- Report Distribution Schedules
- Repot Distribution Lists
- Resource Startup/Shutdown Procedures
- Service Owner Contact List
- Service Startup/Shutdown Procedures
- Shift Turnover Reports
- Shift Turnover Reports
- Spare Parts Inventories
- Staffing Schedules
- Supply Inventories
- Supply Inventory Counts
- Supply Inventory Lists

- Supply Reorder Points
- Tape and Media Archives
- Tape Vault List
- UPS Startup/Shutdown Procedures
- Work Instructions

Policy Database

Holds policies documentation used to govern the support and delivery of IT services.

Logical CIs represented in this database include:

- Change Policy
- Clock/Time Management Policy
- Configuration Control Policy
- Data Retention Policy
- Incident Escalation Policy
- Incident Prioritization Policy
- Information and Document Classifications
- Intellectual Property (ICAP) Policies
- IT Service Continuity Invocation Policy
- Patch Management Policy
- Privacy Policies
- Quality Control Plan
- Release Policy
- Security Incident Response Plan
- Security Policies
- Security Policy
- Security Requirements

Problem Management Database

Holds data related to problem records and the status of problems that have occurred in the IT infrastructure.

Logical CIs represented in this database include:

- Customer Communication Notices
- Customer Statements For Major Incidents
- Problem Status Reports
- Problem Tickets

Procurement Database

Holds data related to third party suppliers, supplier costs and their services.

Logical CIs represented in this database include:

- Network Carrier Service Agreements
- Office and Supply Purchase Receipts
- Service Proposals
- Software License Obligations
- Supplier Catalogs and Brochures
- Supplier Contact Lists
- Supplier Contact Lists
- Supplier Cost Estimates
- Supplier Descriptions
- Supplier Quality Metrics
- Supplier Quality Reports
- Trusted Partners
- Underpinning Contract (UC)
- Vendor Catalogs
- Vendor Leases
- Work Orders

Project Management Database

Holds data related to projects and service improvement programs being undertaken in the infrastructure.

Logical CIs represented in this database include:

- Project Earned Value Analysis Reports
- Project Issue Lists
- Project Portfolio
- Project Status Reports
- Project Templates
- Project Work Plans
- Project Workbooks
- Project Working Standards
- Scope Statements
- Service Improvement Plans (SIPs)
- Work Breakdown Structures

Release Management Database

Holds data related to release units, migration plans and testing activities used to transition new or changed services into production operations.

Logical CIs represented in this database include:

- Customer Constraints
- Customer Requirements
- Deployment Schedules and Plans
- Design Principles and Guidelines
- Engineering Specifications
- Event Monitoring Plan/Architecture
- Migration/Deployment Plans
- Production Readiness Review Checklists
- Release Packages
- Service Design Packages
- Service Solution Prototypes and Demos
- Site Implementation Plans
- Site Surveys
- Software Distribution Managers
- Solution Review Findings
- Test Results
- Use Cases

Request Management Database

Holds data related to requests and their status being fulfilled in the IT infrastructure.

Logical CIs represented in this database include:

- Request Fulfillment Procedures
- Request Status Reports
- Request Tickets
- Service Requests

Security Management Database

Holds data related to security plans, activities, configurations and reports used to protect assets and services in the IT infrastructure.

Logical CIs represented in this database include:

- Access Control Lists
- Access Permissions
- Access Profiles
- Anti-Virus Patches and Downloads
- Badge Control Logs
- Badges
- CCTV Security Camera Video Archives
- Denial Of Service Testing Software
- Door Lock Combinations
- Encryption Schemes
- External Security Interfaces
- Firewall Logs
- Firewall Rules
- Forensics
- Host-based Intrusion Monitors
- Identity Management Applications
- Identity Management Interfaces
- Internet Protocol Security (IPSec) Controls
- Intrusion Detection Log
- Intrusion Detection Reports
- IP Packet Scanners
- Key Lists
- Network-based Intrusion Monitors
- Password Cracking Monitors
- Passwords
- Physical Building Security Event Logs
- Secure Electronic Transmission (SET) Controls
- Secure Hypertext (S-HTTP, HTTPS) Controls
- Secure Shell (SSH) Controls
- Secure Socket Layer (SSL) Controls
- Secured Transmission Logs
- Security Application Interfaces

- Security Audit Control Requirements
- Security Audit Results
- Security Awareness Plan
- Security Escalation Contact Lists
- Security Incident Reports
- Security Keypads
- Security Logs
- Security Patches
- Security Profiles (Matching Patterns)
- Security Test Conditions
- Security Test Results
- Security Vulnerability Assessment
- Single Sign-On (Password Sync) Systems
- Surveillance Logs
- Unauthorized Access Reports
- User IDs
- Virtual Private Network Configurations (VPNs)
- Visitor Access Control Logs
- Visitor Control Logs

Service Catalog Database

Holds data related to service descriptions and software systems used to manage them.

Logical CIs represented in this database include:

- Service Catalog Systems
- Service Descriptions

Service Level Management Database

Holds data related to service level agreements and service reporting systems used in the IT infrastructure.

Logical CIs represented in this database include:

- Operational Level Agreement (OLA)
- Service Level Agreement (SLA)
- Service Reporting Systems

Strategy Database

Holds data related to service and governance strategies used to guide, steer and control support and delivery of IT services.

Logical CIs represented in this database include:

- Communications Plan
- Governance Strategy
- Service Portfolio
- Service Sourcing Strategy

System Directory Database

Holds data related to user IDs, assets and access permissions to them.

Logical CIs represented in this database include:

- Directory Interfaces
- Directory Management Software
- Directory Names
- Directory Permissions
- Directory Structures

Test Management Database

Holds test data, testing plans and test results related to testing services, releases, processes and technology assets used in the IT infrastructure.

Logical CIs represented in this database include:

- Service Test Plans
- Test Data Files
- Test Plan
- Testing Schedules

Training Database

Holds data related to training plans, materials, schedules and delivery requirements.

Logical CIs represented in this database include:

- Training Attendance Status Tracking
- Training Materials
- Training Plans
- Training Schedules
- Training Setup Requirements

Chapter

5

Organization Architecture

Organization Architecture Overview

This chapter provides an organizational view of the IT Service Management architecture. In order to be effective, IT cannot just implement processes and tools. IT needs to organize itself to successfully support and deliver services. Presented here is a comprehensive set of job functions and role descriptions needed to operate an entire IT Service Management Architecture.

The role descriptions presented in this chapter are generic. A role is a set of related activities grouped together. Roles are mapped into job functions. A job function may consist of one or more roles. It is also possible to have multiple job functions fill in a single role.

The means for building an IT service organization vary greatly between business organizations. These range from political concerns to people-centric methods.

Organization Models

One of the key decisions any IT organization has to make is that of what kind of organizational reporting hierarchy needs to be put into place. The reporting hierarchy is essentially needed to enforce proper controls as activities are carried out to deliver services.

The decision is a careful one of balance. Put too much of a hierarchy in place and the organization creates bottlenecks for getting decisions made and accomplishing tasks. Put too little of a hierarchy in place and eventual chaos could result.

The following lists several kinds of general organizational models that can be used for ITSM. They are:

- Centralized Organizational Model
- Distributed Organizational Model
- Networked Organizational Model
- Localized Organizational Model
- Virtualized organizational Model

These can be used singly or in combinations. The following pages describe these in more detail and present pros and cons for using them.

Centralized Organization Model

In this approach, everything reports up through a highly hierarchical structure. Centralized models are highly efficient at executing IT services. This model works especially well with smaller IT organizations. It can be shown as follows:

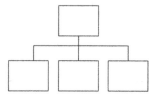

Figure 5: Centralized ITSM Organization Model

With the Centralized Organization Model, most ITSM decision points exist at the upper levels of the hierarchy with lower levels only empowered to carry out those decisions. The downside is that this model can create bottlenecks as the organization grows larger, becomes highly distributed and/or the pace of business and IT change increases.

Positive reasons to use this model include:

- Good for small companies
- No management needed at Local Sites
- Efficient decision making
- Provides high degree of control
- Lower costs

Negative reasons to use this model include:

- Could become bottleneck in larger organizations
- Stifles new ways of thinking
- Slower to adapt to local business needs

Distributed Organization Model

In this approach, major decisions and control points are established centrally at higher levels in the hierarchy. Minor decisions are handled at lower levels as long as they do not violate the goals or principles established centrally. This is pictured as follows:

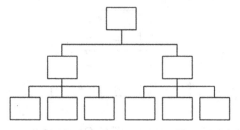

Figure 6: Distributed ITSM Organizational Model

With the Distributed Organizational Model, decisions and control policies are placed centrally with some autonomy at local levels. For this reason, this approach is highly preferred for larger organizations, especially those with many locations.

Positive reasons to use this model include:

- Good for global or large companies
- Management of ITSM solutions is closer to those that execute them
- Still provides high degree of control
- Allows for quick adaptation to local business needs

Negative reasons to use this model include:

- Enterprise standards and control needed to ensure control is not lost
- More expensive to use due to redundant roles at each lower level

Networked Organization Model

This approach can be utilized when there is no central ITSM decision making authority that is being put into place. This is typically seen in situations where many company acquisitions have taken place but they still need to operate as independent autonomous business units. This is pictured as follows:

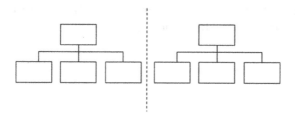

Figure 7: Networked ITSM Organizational Model

With the Networked Organizational Model, there is no central point for decisions. The decision making authority rests within each organizational unit. Note that this creates a much greater risk of variation in processes, technology choices and how services are delivered.

Positive reasons to use this model include:

- Works where a single model cannot be applied
- Good for global or merging companies where autonomy is to be maintained
- Provides management efficiency at local sites

Negative reasons to use this model include:

- Redundancy of roles at all organizational levels
- Much higher costs due to redundancy of positions
- Replication and variation of process is a risk with this kind of organizational model since there is little overall control

Localized Organization Model

This approach can be utilized to establish an ITSM organization at peer level with the business and/or IT leadership level. Using this approach, key ITSM roles such as Process Owners and Service Managers exist with this organization which then guides and directs the existing IT functions. With this, the ITSM organization is *accountable* for ITSM services and processes. The existing IT line functions are *responsible* for running those functions day-to-day. This is pictured as follows:

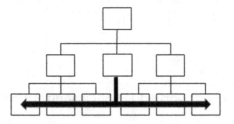

Figure 8: Localized ITSM Organizational Model

Many organizations have adopted this kind of approach successfully with ITSM as it is the least disruptive to the current organization.

Positive reasons to use this model include:

- Provides accountability for both processes and services
- Minimizes disruption to existing IT functions while ITSM practices are taking hold
- Allows each IT unit to focus on their core service delivery and technology capabilities
- Lowers risk of process variance

Negative reasons to use this model include:

- Process and service issues when they arise will have to be addressed at CIO level or through a service governance council

- Must work through cross matrix issues as the ITSM unit works with other IT units to resolve conflicts in priorities

Virtualized Organization Model

This approach is best utilized for IT organizations with existing structures that are very large and complex. With this approach, IT support staff report to two masters: their line function management and ITSM management. Both ITSM functions and line functions exist as virtual organizations sharing IT support staff resources. This is pictured as follows:

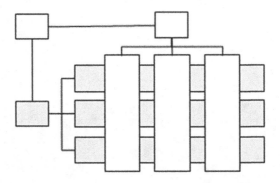

Figure 9: Virtualized ITSM Organizational Model

This approach has been used in very large IT organizations (typically over 10,000 support staff) where reorganization of function is complex and hard.

Positive reasons to use this model include:

- Provides accountability for both processes and services in a very large IT organization
- Allows for complex decision making and prioritization processes
- Forces cooperation in service delivery across each business unit

Negative reasons to use this model include:

- Potential for conflicts between responsibilities for service management versus individual organization unit priorities

- Will still require overall leadership and coordination across units at IT leadership level (or through a selected leader)

Transforming To an ITSM Organization

Not many IT organizations have the luxury of building their organizational solution from scratch. It is very common to find that IT organizations have historically evolved into organizational functions that are heavily siloed by technology platforms and applications. These might look like the following:

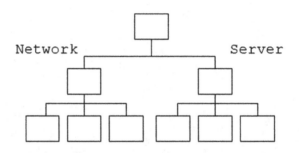

Figure 10: Typical IT Organization With Technology Silos

In the above, each technology silo operates independently to deliver pieces and parts of services. There is no service ownership. The IT director or CIO is essentially the service integrator to make sure overall service needs are being taken care of. Usually this individual wonders why they have to be on the phone to so many IT people when serious outages occur and complain that IT just can't seem to work together.

So when faced with an IT organization organized by technologies, how does one go about transforming it to one that will operate by the services it delivers? The Localized Organizational Model discussed earlier provides an answer:

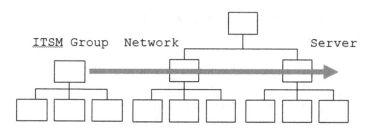

ITSM Group   Network   Server

Figure 11: Suggested ITSM Localized Organization Strategy

With this approach, one additional organizational silo is created devoted to IT Service Management. Inside this group may include roles such as Process Owners, Service Owners, Business Relationship Managers and other key ITSM Continual Service Improvement roles.

These roles then work across all the other silos in a cross functional relationship manner to instill IT Service Management practices. It is critical that this group report directly to the IT director or CIO to make sure that enough authority is given it to make their efforts successful.

At a very minimum, this new organization should include at least Service Owners, Process Owners and Business Relationship Managers. These are essential roles needed to begin transforming the organization from a technology based one to a service based one.

One of the key decisions any IT organization has to make is that of what kind of organizational reporting hierarchy needs to be put into place. The reporting hierarchy is essentially needed to enforce proper controls as activities are carried out to deliver services.

One of the benefits of this approach, gained almost immediately, is that there is now accountability for services, processes and customers in the IT organization where it may not have existed previously.

A more disciplined approach to building this organization can consist of the following 7 steps:

1) Identify the processes and procedures needed.
2) Identify a set of generic roles
3) Map processes/procedures into the roles
4) Assign responsibilities for each process by role
5) Identify skill sets needed to operate each role
6) Identify level of skills needed for each role
7) Group (or map) roles into job functions

The above approach has several distinct advantages. It cuts through the politics and builds the organization based on actual work that has to be accomplished. It creates very clear job definitions and responsibilities that can be clearly understood. It also greatly supports recruiting and training efforts.

Since not everyone has the luxury of building an entire IT service organization from scratch, you can map your results into your existing IT management structure in step 7. Once done, you may discover a number of roles or tasks left hanging with no job function covering them. You may also help straighten out confusion over roles and responsibilities that may have existed previously.

The following pages illustrate a generic organizational model that can be used for operating an IT Service Management infrastructure. Organization structures can vary greatly from one business to another. There is no one right answer for this, but whatever organization structure is in place, should conceptually align with the models shown here.

The high level organization model for an IT Service Management focused organization can be pictured as shown below:

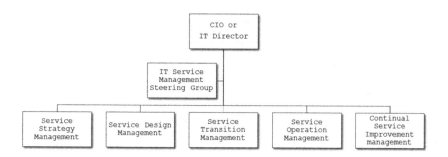

Figure 13: ITSM Suggested Functional Model Executive Level

In the above model, there is a CIO or IT Director overseeing the entire organization. Direct reports represent functional managers covering each stage of the ITSM Service Lifecycle.

A breakdown of the Service Strategy function would look as follows:

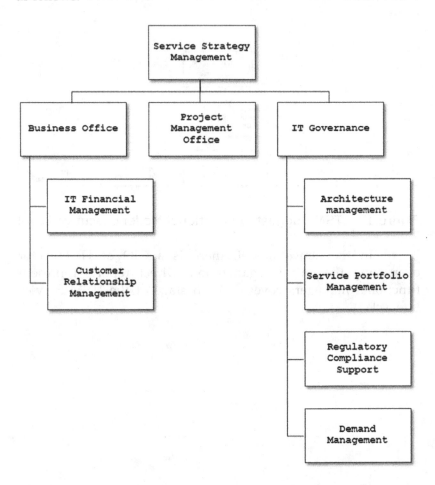

Figure 14: ITSM Suggested Functional
Model Service Strategy

A breakdown of the Service Design function would look as follows:

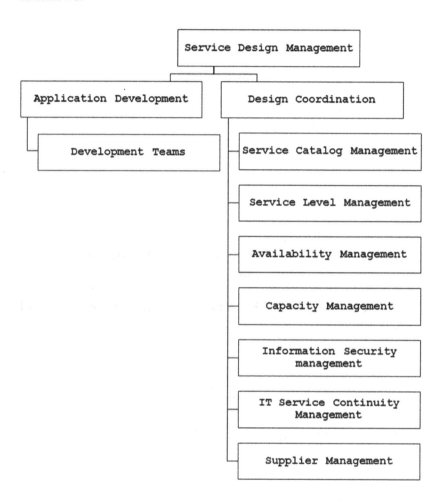

Figure 15: ITSM Suggested Functional
Model Service Design

A breakdown of the Service Transition function would look as follows:

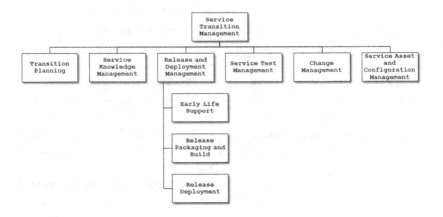

**Figure 16: ITSM Suggested Functional
Model Service Transition**

A breakdown of the Service Operation function would look as follows:

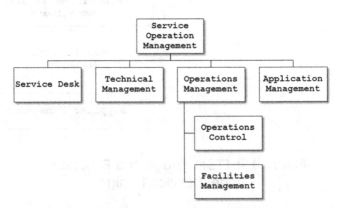

**Figure 17: ITSM Suggested Functional
Model Service Operation**

A breakdown of the Continual Service Improvement function would look as follows:

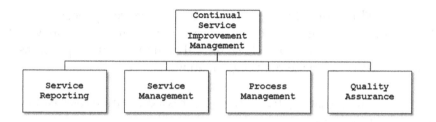

Figure 18: ITSM Suggested Functional
Model Continual Service Improvement

Detailed role descriptions are shown on the following pages. A mapping of those role descriptions to each of the job functions shown in the above models is also included.

General IT Service Management Roles

This set of roles support all service, process and technology activities that are not specific to an ITSM Lifecycle Stage. They represent high level roles that are needed to operate as an IT Service Management organization.

Mappings of job functions addressed by these roles can be shown as follows:

Role	Job Function(s)
IT Executive	CIO/IT Director ITSM Steering Group
IT Functional Unit Manager	Service Strategy Mgt Service Design Mgt Service Transition Mgt Service Operation Mgt Cont. Service Imp. Mgt Business Office Project Mgt Office IT Governance
Subject Matter Expert (SME)	Supports any job function
Steering Group Member	ITSM Steering Group

The Subject Matter Expert (SME) described in this chapter can be ascribed to any job function as that role provides expertise when needed.

IT Executive

This role provides oversight and governance over IT activities. It sets the mission for IT, agrees the IT services and obtains funding needed to meet IT strategic goals. It also acts as the highest authority to resolve IT service issues and make key decisions when needed.

- Provides a leadership role in building and organizing the IT service support and delivery organization.
- Obtains funds for IT Service Management service improvement activities, development of new services and ongoing service operations.
- Provides approval role for the IT budget.
- Provides highest point of escalation for IT decisions and resolution of IT issues.
- Provides a leadership and responsibility role for developing, executing and maintaining the IT strategy.
- Provides official approval for IT investments.
- Provides a visible role as an IT change leader promoting IT Service management culture and discipline.
- Assists and guides negotiations for major IT purchases.
- Assists and reviews business cases for major IT purchases.
- Ensure IT services are compliant with business, legal and regulatory requirements.
- Interacts with executive management teams to monitor and validate the enterprise's compliance with IT security policies.

IT Functional Unit Manager

This role provides oversight and management of one or more IT functional organizational units. It ensures that units are appropriately staffed, manages issues, manages adherence to standards and processes and ensures delivery goals are met.

- Recruits, hires and trains IT staff for functional units managed.
- Manages the functional unit budget.
- Approves purchases for functional unit within corporate guidelines and policies.
- Conducts effective performance evaluations for staff and mentors those with less experience through formal channels.
- Helps staff execute career development plans.
- Suggests areas for improvement in internal processes along with possible solutions.
- Approves staff time and expense reports.
- Reviews the status reports of staff team members, the projects they work on and addresses issues as appropriate.
- Complies with and helps to enforce standard policies and procedures.
- Manages day-to-day internal staff interactions.
- Sets and manages internal stakeholder expectations.
- Communicates effectively with stakeholders to identify needs and evaluate alternative solutions.
- Continually seeks opportunities to increase stakeholder satisfaction and deepens relationships.

Subject Matter Expert (SME)

This role provides expertise in technical, business, operational and/or managerial areas when requested.

- Provides technical, operational, business and/or managerial subject matter expertise.
- Provides input into the design of the procedures, tools or organization as required.
- Assists in the development of ITSM solutions by providing specialized expertise as required.
- Supports the development and execution of test scenarios designed to validate the functionality of the design.
- Validates the Design and Implementation Team designs for processes, tools and organization and any recommendations.
- Provides consultative and facilitation support to the Implementation Project Teams.
- Assists in creation of project work plans and implementation strategies.
- Provides Intellectual Capital as required during the Implementation Project.
- Coaches team members in specialized skill sets if required.

Steering Group Member

This role provides steering oversight over IT strategies, mission and services. It ensures that IT priorities and plans are in alignment with business priorities and plans.

- Champions IT Service Management solutions and practices across the enterprise.
- Conducts periodic meetings to monitor ITSM services progress and issues.
- Provides final review and approval of continual service improvement project deliverables.
- Coordinates approvals from business units as necessary.
- Identifies and appoints key ITSM staff and project team members.
- Coordinates major program decisions that have been escalated to the Steering Group on a timely basis to meet service objectives.

Service Strategy Roles

This set of roles support the ITSM Service Strategy Lifecycle stage. Mappings of job functions addressed by these roles can be shown as follows:

Role	Job Function(s)
IT Financial Manager	IT Financial Management
IT Financial Analyst	IT Financial Management
IT Financial Administrator	IT Financial Management
Demand Manager	Demand Management
IT Service Portfolio Analyst	Service Portfolio Management
IT Market Analyst	Service Portfolio Management
Project Manager	Project Management Office

IT Financial Manager

This role is responsible for managing activities that provide cost effective stewardship of IT service assets used to provide IT services.

- Develops IT account plans and investment cases when needed by organizations and business units.
- Manages the IT organization budget.
- Prepares IT budget forecasts and assist other organizations, when necessary, in preparing the IT elements of their budgets.
- Reports regularly on budget conformance to IT managers.
- Identifies budget conformance issues to IT managers.
- Provides close support to Service Level, Availability, Capacity and IT Service Continuity management processes during budgeting and IT investment planning.
- Produces IT financial plans in line with business planning cycles, identifying financial needs early enough to accommodate procurement and approval lead times.
- Participates in CAB meetings to assess and authorize changes from an IT Financial perspective.
- Manages and operates IT billing and chargeback operations.
- Resolves chargeback and billing disputes when they arise.
- Conducts periodic internal audits of IT financial performance.
- Assists with external audits of IT financial performance when requested by other corporate organizations.
- Ensures IT charges and billings are accurate and truly reflect the services delivered.

IT Financial Analyst

This role supports IT Financial Management activities with analysis of IT costs, budgets and chargeback operations.

- Advises Service Level Manager about IT costs and charges for service levels or service level options.
- Identifies suitable tools and processes for gathering IT cost data.
- Develops suitable IT cost models.
- Assists in developing cost/benefit cases for IT investments.
- Advises management on the cost-effectiveness of IT solutions.
- Assists with external audits when requested.
- Provides support for IT service charging activities when necessary including:
    - Identification of charging policies
    - Providing justifications and comparisons for charges
    - Preparing regular bills for customers
    - Preparing price lists for services
- Conducts ad-hoc performance and IT Financial studies on request from IT management.
- Analyzes and breaks down IT infrastructures into cost components and categories.
- Builds and develops IT cost models.
- Communicates the impacts of planned IT investments.

IT Financial Administrator

This role provides administrative support to handle tasks that gather cost and budget data, administer financial reporting tools, maintain cost models, assemble and produce financial reports and any other tasks that assist the IT Financial Management process.

- Administers IT Chargeback and billing operations.
- Gathers cost and budget data when requested.
- Assembles budget, accounting and charging reports.
- Maintains Financial Management database.

Demand Manager

This role identifies patterns of business activity that drive the usage of IT services, tracks and forecasts that activity, controls that activity through financial means when necessary.

- Identifies, analyses, and codifies business activity patterns to identify demand factors that impact capacity and use of IT service assets.
- Maintains inventory of demand factors and their associated business volumes.
- Reviews demand factors and business volumes on an ongoing periodic basis with key business staff and agrees forecasts for these volumes.
- Interfaces with capacity management process to develop capacity workload characterizations for demand factors identified.
- Interfaces with IT Financial Management processes to control and manage demand via service pricing policies.
- Works with Capacity management to model future impact of business decisions and plans by forecasting anticipated changes to demand factors and business volumes.

IT Service Portfolio Analyst

This role ensures that the contents of the IT Service Portfolio are accurate and kept up to date. It gathers up necessary data to maintain the portfolio and works with senior executives to keep the portfolio viable.

- Builds, maintains and reports on a portfolio of IT service investments based using risk and reward profiles.
- Assist IT executives in maximizing return on their IT investment portfolio by keeping the portfolio current and viable.
- Assist IT executives in prioritizing and aligning portfolio initiatives and priorities based on changing business, market and regulatory needs.
- Maintain viability and status of the IT service pipeline (services under development) and its impact on current investments and priorities.
- Assist IT executives in identifying and analyzing level of customer value for each service in the portfolio.
- Assist IT executives in identifying portfolio investments in IT services that may no longer be providing value.
- Assist executives in identifying impacts to the IT Service Portfolio from business decisions, plans and strategies.

IT Market Analyst

This role performs market research within particular customer industry segments for IT services to determine accurate market descriptions, market trends, forecasts and models. It also markets the IT Service Catalog to business unit customers and stakeholders.

- Conducts efforts to deliver a combination of market research, competitive intelligence, and management consulting to business units and executive staff.
- Surveys and interviews IT technology and service providers and their key supply chain partners, technology buyers and channel partners to obtain industry needs and trends.
- Analyzes business unit and stakeholder customer needs to help identify business problems and propose potential IT service solutions.
- Communicates results of market research and analysis to IT executives.
- Identifies options for improving IT services and bridging the needs of the business with the use of IT.
- Markets IT Service Catalog and the services it contains to business units and other IT stakeholders.
- Identifies customer perception of value and desired outcomes.
- Assists in identifying functional requirements for new or changed IT services.

Project Manager

This role provides oversight and management of IT projects. It manages project activities on a day to day basis and reports project status to IT executives.

- Responsible for assigned project objectives.
- Provides direction to the project teams for work products due as well as the overall status of the projects assigned.
- Co-ordinates activities with other project managers when necessary.
- Provides status of work in progress and/or issues to the Program Manager
- Develops project work plans, schedules and staffing requirements for projects assigned.
- Communicates as required to executive management or Program Office staff.
- Conducts weekly change, issues and status meetings to track progress and risks with Core Teams assigned to.
- Ensures that outstanding project management, process implementation and design requirements and/or issues are being addressed for projects assigned.
- Escalates cross project issues or key management issues to the Program Manager.
- Communicates activities and status of each project assigned throughout its lifecycle.
- Schedules workshops and meetings as required.
- Provides overall leadership and management for the projects assigned.

Service Design Roles

This set of roles support the ITSM Service Design Lifecycle stage. Mappings of job functions addressed by these roles can be shown as follows:

Role	Job Function(s)
Solution Architect	Architecture Management
Service Catalog Administrator	Service Catalog Management
Service Level Manager	Service Level Management
Service Level Analyst	Service Level Management
Availability Manager	Availability Management
Availability Analyst	Availability Management
Availability Architect	Availability Management
Capacity Manager	Capacity Management
Capacity Analyst	Capacity Management
Capacity Architect	Capacity Management
ITSCM Manager	IT Service Continuity Management
ITSCM Analyst	IT Service Continuity Management
ITSCM Team Member	Can apply to any job area
Chief IT Security Officer	Regulatory Compliance
IT Security Manager	Information Security Management
IT Security Analyst	Information Security Management
IT Security Auditor	Information Security Management
Supplier Manager (Contract Manager)	Supplier Management
Supplier Liaison	Supplier Management

Solution Architect

This role establishes the overall architecture for service solutions. It also coordinates common technology related activities between all project teams involved in the service solution design effort.

- Ensures the solution architecture meets the strategic needs of business requirements.
- Coordinates solution product selections and tailoring.
- Supports cross team early launch planning from a technology perspective.
- Ensures maximum integration of tools and technologies that underpin solutions.
- Coordinates technology solution implementation activities.
- Coordinates technology customization and integration activities.
- Coordinates Technical resources to optimize use of technology solutions.
- Identifies ongoing support and maintenance for technologies chosen.
- Communicates chosen solution architectures to program teams.
- Interfaces to technology vendors as needed.

Service Catalog Administrator

This role is responsible for ensuring that the IT Service Catalog is accurate, up to date and reflective of the IT service strategy.

- Documents and administers the IT Service Catalog.
- Ensures IT Service Catalog accurately reflects the IT Service Portfolio and strategies.
- Publishes and communicate the IT Service Catalog throughout IT and the business.
- Responds to RFCs for service changes and additions by updating the catalog as needed.
- Ensures service descriptions in the catalog are accurate and inclusive of all service features designed.
- Publishes information to obtain and receive help and support for services published in the catalog.
- Ensures catalog includes descriptions for upcoming services about to be offered as well as services planned for retirement.
- Maintains historical versions of published catalogs.
- Manages versions and revisions of the IT Service Catalog as needed.

Service Level Manager

This role defines, negotiates, monitors and reports service level agreements (SLAs), Operational Level Agreements (OLAs) and Underpinning Contracts (UCs). It makes recommendations for service improvements and performs regular reviews to compare delivered service quality to service commitments.

- Collects SLA, OLA and Underpinning Contract performance metric data.
- Defines, documents, negotiates, agrees and publishes service level agreements (SLAs).
- Develops and negotiates OLAs and Underpinning Contacts and ensure they meet defined Service Level Agreements.
- Works with Service and Process Owners to identify service improvement projects as needed to correct or improve services.
- Negotiates service level commitments with business units and customers.
- Provides general SLA oversight.
- Provides oversight and effectively manages OLAs and UCs.
- Responds to RFCs for changes to SLAs, OLAs or Underpinning Contracts.
- Ensure service agreements are reflected in the IT Service catalog and well communicated to customers and stakeholders.
- Provide support for identifying needed service targets for agreements and supplier contracts.
- Review agreements and report on their attainment and trends

Service Level Analyst

This role provides administrative support to handle tasks that gather service data, administer service reporting tools, assemble and produce service reports and any other tasks that assist the Service Level Management process.

- Administers Service Level Management support tools.
- Assembles and produces Service Level Management reports.
- Handles ad-hoc requests for Service Level Management status or one-time information retrieval/reports.
- Manages service reporting communication distribution lists.
- Gathers service data as needed for reporting and communications.

Availability Manager

This role has maintains the availability and reliability of IT services to ensure that IT can effectively meet service targets in accordance with planned business objectives. It identifies and specifies availability business, instrumentation/measurement, new monitoring, and supplier requirements.

- Optimizes the availability of the IT infrastructure to deliver cost effective improvements that deliver tangible benefits to business units and customers.
- Provides holistic management of availability that includes people and processes as well as technology.
- Takes actions to achieve reductions in frequency and duration of incidents that impact IT availability.
- Ensures shortfalls in IT availability are recognized and appropriate corrective actions are identified and progressed.
- Creates and maintain a forward looking availability plan aimed at improving the overall availability of IT services and infrastructure components to ensure that existing and future availability requirements can be met.
- Provides regular reports on availability to the Service Manager.
- Organizes and maintain a regular availability review process with both key business and IT representatives.
- Agrees appropriate actions to maintain or improve availability levels.
- Initiates and coordinates actions required to maintain or improve availability across IT units.
- Acts as a coordination point for changes to availability levels when needed.
- Undertakes risk assessments to identify single points of failure or other areas in the infrastructure that place availability of services at risk.
- Works with the business to identify vital business functions requiring IT support that are critical to business operations.

Availability Analyst

This role analyzes existing availability issues and problems to determine ways to improve availability at acceptable cost levels and determine availability requirements for new IT solutions and service changes.

- Provides a range of IT availability reporting to ensure that agreed levels of availability, reliability and maintainability are measured and monitored on an ongoing basis.
- Determines the availability requirements from the business for new or enhanced IT services.
- Establishes measures and reporting that reflect business, user and IT support organization requirements.
- Monitors actual availability achieved versus targets and to ensure shortfalls are addressed.
- Participates in Change Control meetings to assess and authorize changes from an availability perspective.
- Conducts availability risk assessment for existing services.
- Assists in SLA negotiation efforts from an availability capability standpoint.
- Defines the key targets of availability required for the IT infrastructure and its components that underpin a new or enhanced IT service as the basis for an SLA agreement.
- Analyzes and reviews actual availability levels achieved against SLAs and OLAs and UCs.
- Maintains an awareness of technology advancements and best practices that support availability.
- Gathers availability data as needed for reporting and communications.

Availability Architect

This role assists Availability Management initiatives by providing overall strategy and design direction.

- Creates availability and recovery design criteria to be applied to new or enhanced infrastructure design.
- Ensures IT services are designed to deliver the required levels of availability required by business units.
- Ensures the levels of IT availability required are cost justified.
- Assists Business Relationship Managers in identifying availability needs for business units.
- Documents availability blueprints and designs as needed.

Capacity Manager

This role ensures that adequate IT capacity exists to meet required levels of service and for ensuring that IT management is correctly advised on how to cost effectively match capacity and demand. It develops Capacity Plans, cost summaries and refresh requirements/estimates. It develops capacity requirements to meet future business requirements.

- Produces capacity plans in line with business planning cycles; identifying requirements early enough to accommodate procurement and approval lead times.
- Documents need for increases and reductions in hardware based on service level requirements, targets and cost constraints.
- Provides regular management reports which include current usage of resources, trends and forecasts.
- Coordinates and oversees performance testing of new systems and solutions.
- Provides holistic management of capacity that includes people, processes and technology.
- Oversees decisions and actions to utilize Demand Management for controlling capacity when necessary.
- Ensures shortfalls in capacity are recognized and appropriate corrective actions are identified and progressed on a timely basis.
- Creates and maintains capacity forecasts to predict future hardware/software spend.
- Provides regular reports on capacity to the Availability Manager.
- Organizes and maintains a regular capacity review process.
- Agrees appropriate actions to maintain or improve capacity levels.
- Initiates and coordinates actions required to maintain or improve capacity across IT units.
- Acts as a coordination point for changes to capacity levels when needed.

Capacity Analyst

This role analyzes capacity issues and problems to determine ways to improve capacity and performance at acceptable cost levels.

- Advises the Availability Management process about appropriate service levels or service level options based on capacity capabilities.
- Sizes all proposed new systems to determine computer and network resources required taking into account hardware utilizations, performance service targets and cost implications.
- Reports on performance against service targets contained in Service Level Agreements.
- Maintains a knowledge base of future demand for IT services and predict the effects of demand on performance targets and service levels.
- Models impacts of changes in business volumes and new technologies to predict needed capacity.
- Translates business events and drivers into IT workloads and volumes.
- Determines performance targets and service levels that are achievable and cost justified.
- Conducts ad-hoc performance and capacity studies on request from IT management.
- Ensures requirements for reliability and availability are taken into account in all capacity planning and sizing activities.
- Analyzes and reviews actual Capacity levels achieved against SLAs and OLAs and UCs.
- Provides a range of IT capacity reporting to ensure that agreed levels of capacity and performance are measured and monitored on an ongoing basis.
- Determines the capacity requirements from the business for new or enhanced IT services.
- Monitors actual capacity usage versus targets and to ensure shortfalls are addressed.
- Participates in CAB meetings to assess and authorize changes from a capacity perspective.

- Conducts capacity risk assessments when needed.
- Assists in SLA negotiation efforts from a capacity capability standpoint.
- Defines the key targets of capacity required for the IT infrastructure and its components that underpin a new or enhanced IT service as the basis for an SLA agreement.
- Maintains an awareness of technology advancements and best practices that have impact on capacity.
- Gathers capacity data as needed for reporting and communications.

Capacity Architect

This role assists in Capacity Management initiatives by providing overall strategy and design direction.

- Recommends resolutions to performance related incidents and problems.
- Creates capacity and performance design criteria to be applied to new or enhanced infrastructure design.
- Ensures IT services are designed to deliver the required levels of capacity and performance required by business units and customers.
- Ensures the levels of IT capacity required are cost justified.
- Identifies capacity needs for the Business Units represented to IT.
- Documents capacity related blueprints and designs as needed.

IT Service Continuity Manager

This role puts plans and strategies in place to ensure that IT services can be recovered in the event of a major business disruption. It maintains recovery plans and oversees ongoing testing, awareness, education and plan maintenance activities.

- Ensures that IT service continuity plans are kept current and up to date.
- Maintains IT Service Continuity planning documents.
- Develops and manages IT Service Continuity plans to ensure that agreed recovery objectives can be achieved.
- Ensures that all IT Service areas are prepared and able to respond to an invocation of the Continuity Plan.
- Maintains a comprehensive testing schedule.
- Participates in negotiation and management activities with providers of third party recovery services when necessary.
- Participates in CAB meetings to assess and authorize changes from an ITSCM perspective and ensure proposed changes do not compromise continuity plans.
- Provides regular reports on continuity readiness, plan test results and other related issues to the IT Service Level Manager.
- Initiates any actions required to maintain or improve IT service continuity.
- Develops and coordinates IT Service Continuity awareness, testing, training and communication activities.

IT Service Continuity Team Leader

This role represents an IT or business unit by providing continuity requirements and acting as a coordination point for conducting recovery actions in the event of a major business disruption. It also represents a single point of contact for those areas represented during plan testing as well as recovery actions if plans are invoked.

- Acts as a single point of contact to coordinate requirements on behalf of the department or business unit being represented.
- Acts as a single point of contact to coordinate recovery actions in the event of a major business disruption on behalf of the department or business unit represented.
- Maintains current skills and knowledge in recovery actions as dictated by the IT Service Continuity Plan.
- Provides input to the IT Service Continuity Plan on behalf of the organization represented.
- Participates in IT Service Continuity training activities.
- Participates in IT Service Continuity awareness and communication activities when requested.
- Coordinates and integrates IT Service Continuity activities with Business Continuity Management activities at the business unit/department level.

IT Service Continuity Team Member

This role assists with IT Service Continuity recovery actions in the event of a major business disruption as dictated by the IT Service Continuity Team Leader.

- Ensures contact information is kept current and accurate.
- Executes recovery activities as dictated by the IT Service Continuity Team Leader in the event of a major business disruption.
- Assists in IT Service Continuity testing activities if requested.

Chief IT Security Officer

This role is responsible for the development and implementation of IT Security Policies and Standards. It oversees and coordinates activities to maintain confidentiality, integrity and availability of all IT service assets and data.

- Develops and maintains IT Security Policies, Processes and Standards.
- Ensures IT Security functions are carried out in compliance with corporate security guidelines and objectives.
- Provides guidance and assistance to new IT initiatives and projects to ensure appropriate levels of security are included with new solutions.
- Ensures IT Security functions are carried out in compliance with legal and statutory obligations.
- Designs and oversees development of Security Management infrastructure.
- Liaisons with customers on security requirements and Incident Response.
- Reviews OLAs, SLAs to ensure compliance with Security Management goals and objectives.

IT Security Manager

This role supports the development and implementation of IT Security Policies and Standards. It manages all security related projects, oversees security staff and reports on security vulnerabilities and incident status.

- Provides timely reporting and escalation of security incidents.
- Oversees actions taken to resolve major security incidents.
- Ensures vendor provided security maintenance is appropriately applied to IT assets and infrastructure.
- Conducts regular and periodic audits and tests of IT security.
- Oversees production of regular and periodic security reports.
- Acts as a focal point for special security investigations when requested.
- Oversees the SOC (Security Operations Center).
- Recommends changes and improvements to Security Management support tools.

IT Security Analyst

This role provides support for Security Management functions by designing security infrastructure solutions and ensuring solutions align with security policies and standards.

- Installs and administers security tools.
- Assists with communications of security-related policies and information.
- Monitors infrastructure for security violations and incidents.
- Logs all security incidents.
- Supports Incident and Problem Management activities from a security perspective.
- Assists in the resolution of security incidents and problems.
- Assists with security audits when requested.
- Assists with security testing when requested.

IT Security Auditor

This role provides is responsible for audit support and testing of the Security infrastructure.

- Conducts regular and periodic tests of the security infrastructure.
- Identifies security test criteria.
- Maintains the security testing plan.
- Participates in external security audit activities when requested.
- Ensures security infrastructure is compliant with corporate, legal and statutory security guidelines and requirements.
- Identifies security audit and testing issues when they arise and ensure they are being appropriately addressed.
- Estimates security risks in the infrastructure.

Supplier Manager (Contract Manager)

This role oversees all suppliers to IT and ensures that they are meeting their service obligations. It reports on supplier quality, addresses supplier issues and represents IT interests when contracting with suppliers.

- Oversees activities to manage suppliers to ensure they meet provider service targets and objectives.
- Coordinates Supplier Liaison activities.
- Interfaces with business legal functions as needed to support vendor contracting and handle contract issues.
- Oversees activities to negotiate needed services and service objectives with external suppliers.
- Responds to escalated supplier related issues.
- Develops Request for Proposal (RFP) and Request for Information (RFI) documents.
- Oversees supplier selection process.
- Provides results from vendor evaluations.
- Ensures requirements for vendor services are appropriately documented and communicated.
- Reviews reports that provide information on supplier quality and adherence to contracted services.
- Coordinates vendor question and answer sessions during vendor RFP and selection process.
- Supports activities to conduct vendor quality checks such as ensuring vendor has appropriate financial resources to deliver supported services.

Supplier Liaison

This role provides a single point of contact to one or more suppliers of IT services to communicate service needs and requirements and validate that represented suppliers are meeting their service obligations.

- Provides single point of contact for assigned vendors into the IT and business organization.
- Communicates service needs and requirements to assigned vendors.
- Reports on vendor attainment towards contracted services and service targets.
- Escalates vendor related issues to the Supplier Manager.
- Validates that assigned suppliers are meeting their service objectives.
- Assists with RFP, RFI and vendor selection activities as dictated by the Supplier Manager.
- Assists with activities to audit vendors as dictated by the Supplier Manager.

Service Transition Roles

This set of roles support the ITSM Service Transition Lifecycle stage. Mappings of job functions addressed by these roles can be shown as follows:

Role	Job Function(s)
Change Manager	Change Management
Change Administrator	Change Management
CAB Member	Can be any job function
ECAB Member	Can be any job function
Change Owner	Can be any job function
Release Manager	Release and Deployment Management
Release Owner	Can be any job function
Test Manager	Service Testing Management
Testing & Validation Analyst	Service Testing Management
Configuration Manager	Service Asset and Configuration Management
Configuration Analyst	Service Asset and Configuration Management
Configuration Librarian	Service Asset and Configuration Management
Asset Manager	Service Asset and Configuration Management
Asset Administrator	Service Asset and Configuration Management
Procurement Analyst	Release Deployment

Role	Job Function(s)
License Administrator	Service Asset and Configuration Management
Transition Manager	Transition Planning Release Deployment
Release Build Manager	Release Packaging and Build
Deployment Analyst	Release Deployment
Knowledge Manager	Service Knowledge Management
Knowledge Architect	Service Knowledge Management
Knowledge Administrator	Service Knowledge Management
Knowledge Owner	Can be any job function
Trainer	Service Knowledge Management
Training Administrator	Service Knowledge Management
Training Architect	Service Knowledge Management
Organizational Change Leader	Transition Planning
Technical Writer	Service Knowledge Management

Change Manager

This role is responsible for coordinating and managing a structured set of activities to process, coordinate, approve and report on all changes to the IT infrastructure.

- Selects the membership of the Change Advisory Board (CAB), ensuring proper representation.
- Prepares the CAB meeting agenda and facilitates the weekly CAB meeting.
- Updates and distributes the Forward Schedule of Change (FSC) calendar.
- Approves the risk classification determination for a Request for Change (RFC).
- Supports the activities leading to the closure of the RFC.
- Approves RFCs that fall under the Standard Change Models definitions.
- Monitors and reviews the progress of all open RFCs until their completion.
- Monitors process audit results.
- Reviews and track all open RFCs.
- Ensures all RFCs are properly prioritized to meet existing Service Level Agreements (SLA).
- Examines RFCs in jeopardy of meeting SLAs.
- When required, conducts a Post-Implementation Review (PIR) with the Release Manager upon the completion of an RFC.

Change Administrator

This role processes and administers Requests for Change (RFCs) that are submitted to the Change Management process. Activities include receiving, filtering and logging of changes as well as initial analysis of them for impact and communicating changes to appropriate CAB members.

- Support the creation of the RFC.
- Determines the impact, risk, urgency and classification for implementing the change.
- Provides status updates to the Change Initiator.
- Reviews implementation plans and schedules to identify what must be performed to accomplish the change.
- Supports execution of and participates in PIRs to validate the results of the RFC.
- Determines the Change Initiator satisfaction with the change request.
- Supports facilitation of CAB meetings.
- Documents results and reasons for rejection.
- Cancel RFCs.
- Closes RFC.

CAB Member

This role represents a business unit, department, IT function, customers or other areas within IT or the business to review submitted changes and assist the Change Manager in approving and scheduling changes.

- Assists the Change Manager in approving and scheduling changes.
- Ensures that all changes to the technical solution are fully assessed for risk, impact, funding, and funding approval.
- Approves, rejects, or requests further analysis on all RFCs after full consideration of the information available.
- Prioritizes approved RFCs and updates the FSC calendar once scheduling conflicts have been resolved.
- Reviews and approves test and back-out plans for change implementation.
- Reviews and may participate in change Post Implementation Reviews (PIRs).

ECAB Member

This role represents a business unit, department, IT function, customers or other areas within IT or the business to review submitted changes and assist the Change Manager in approving and scheduling changes.

- Reviews emergency changes as appropriate and provide details of their likely impact, the implementation resources, and the ongoing costs.
- Is available for consultation should an emergency Change be required.

Change Scheduler

This role schedules changes and maintains the Forward Schedule of Change calendar. Scheduling activities are performed at the direction and approval of the Change Manager.

- Maintains the Forward Schedule of Changes (FSC) calendar.
- Communicates the FSC to all change stakeholders.
- Communicates desired calendar changes from stakeholders back to the Change Manager.
- Confirms that scheduled changes actually took place.

Change Owner

This role is the single point of contact on behalf of the requestor for submitting a change, communicating any additional concerns or requirements and validating that the change was implemented successfully.

- Provides single point of contact on behalf of the business or IT unit for a requested change.
- Provides authoritative conformation that a requested change was implemented successfully or its final state if not.
- Coordinates communications between those requesting changes and IT as needed.
- Escalates issues and obtains decisions in a timely manner as needed from the business unit requesting the change.
- Reports status of change represented back the business units that requested it.

Release Manager

This role oversees, manages and coordinates all activities to identify releases, develop release plans, assign releases to appropriate IT staff, report on progress of releases, and coordinate release activities with the Change and Service Asset and Configuration Management processes.

- Receives logs, qualifies and assigns all Release requests.
- Coordinates information with Transition Manager and Program Office Manager
- Assigns requested Changes to specific Releases
- Identifies the Appropriate Template
- Conducts Release plan reviews
- Updates Existing Release Plans
- Updates Master Release Schedule
- Interfaces with Change Scheduler to schedule release implementations
- Schedules Release readiness reviews
- Works closely with Change management and participates in CAB meetings
- Reports on status of Releases
- Participates in Post Implementation Reviews (PIRs)
- Escalates Release

Release Owner

This role takes a holistic view of an IT infrastructure change considering all aspects of that change that are both technical and non-technical. It acts as a single point of contact for any given release, develops the release implementation plans and coordinates all activities involved with planning, building, testing and deploying the release.

- Develops Integrated Release and Test Plans
- Assembles submission package for a release
- Initiates Procurement
- Assigns Resources for Release
- Coordinates Rollout Planning
- Develops Implementation Plan, Communication Plan, and Training Plan
- Manages Logistics for Release
- Updates Problem / Known Error(s)
- Ensures requirements for a Release have been provided and provides documentary proof
- Acts as single point of contact for a specific Release
- Coordinates Release Implementation and Transition to Production
- Executes Communication Plan
- Bears responsibility for verifying that the Release has been implemented successfully
- Determines Remedial Action
- Requests Problem ticket based on Post-Implementation Test
- Closes Release
- Escalates key release decisions to executive management

Test Manager

This role oversees, manages and coordinates all activities to ensure structured means are used for testing releases. This includes development and oversight of test labs, testing methodologies, scheduling of testing resources, and maintenance of testing tools and databases.

- Defines and Builds Release Testing Strategy
- Oversees coordination and development of test lab facilities
- Schedules use of test lab resources
- Oversees test lab management staff
- Reviews Releases and assigns appropriate Release testing tasks
- Identifies Release Test Cases
- Identifies Expected Testing Results
- Compiles and Reviews the Testing Deliverables
- Conducts supporting documentation review
- Conducts Release test reviews
- Documents and Publishes Test Results

Testing and Validation Analyst

This role coordinates plans and executes all testing activities on behalf of any release to validate that it is meeting utility, warranty and service requirements. Ideally, this role should be independent of those involved with release design and build responsibilities.

- Conducts installation procedure tests
- Conducts functional testing
- Conducts performance testing
- Conducts integration testing
- Conducts user acceptance testing
- Conducts operational readiness testing
- Conducts back out testing
- Compiles test results
- Forwards test results to the Test Manager

Configuration Manager

This role is responsible for all oversight, planning, maintenance, status accounting and reporting of IT configurations and service models. It reviews accuracy of the Configuration Management System (CMS) and ensures appropriate tools and reporting are in place.

- Oversees development of Configuration Management System
- Plans for Configuration Management databases and activities.
- Identifies Configuration Items.
- Controls Configuration Item information
- Performs status accounting.
- Performs verification and audit of Configuration Management databases.
- Provides management information about Configuration Management quality and operations.

Configuration Analyst

This role identifies new Configuration Items (CIs), and their needed attributes. It assists the Configuration Manager with planning, auditing, validating, status accounting and reporting activities. It identifies any situation that may require modification to configuration structures or design of the CMS.

- Implements Configuration Management System.
- Plans for Configuration Management databases and activities.
- Identifies Configuration Items.
- Controls Configuration Item information.
- Performs status accounting.
- Performs verification and audit of Configuration Management databases.
- Provides management information about Configuration Management quality and operations.
- Sets parameters for the routine maintenance of CIs.
- Associates names with CIs and sets of CIs within the defined schema.
- Initializes, resets, and closes down CIs.
- Collects information about the current state of CIs.
- Obtains announcements of significant changes in the state/condition of CIs.
- Changes the configuration of CIs or sets of CIs.

Configuration Librarian

This role is responsible for maintaining up-to-date (and historical) records of configuration items in the CMS. Activities are performed under the guidance and approval of the Configuration Manager.

- Maintains the Configuration Management Database.
- Responds to requests for CI changes and updates from Change Management.
- Provides CI information upon request.

Asset Manager

This role oversees and manages day to day activities to verify hardware, software and networking asset information and current lease/cost status. It oversees status information for assets from their initial order to delivery, installation, production use and eventual retirement and disposal.

- Plans for Asset Management databases and activities.
- Identifies Asset Items.
- Controls Asset information.
- Performs verification and audit of Asset databases.
- Provides management information about Asset Management quality and operations.
- Communicates asset costs to executives and IT Financial Management.
- Identifies asset issues on a timely basis such as assets coming off lease or invalid software licenses.
- Supports and schedules audits of assets and asset information.

Asset Administrator

This role administers IT asset registration actions, performs asset tagging, and validates asset information and registration. It also processes all requests for verification of IT assets, validates and resolves issues related to missing assets, inaccurate information or status. It acknowledges receipt of assets when delivered and coordinates their return or disposal.

- Maintains the Asset Management Database.
- Records assets when recognized.
- Responds to requests for asset information changes and updates from Change Management.
- Provides asset information upon request.
- Tags and tracks all IT assets, their locations and owners.
- Administers the Asset Management Database and the Asset Inventory.

- Performs asset disposal tasks in line with corporate asset policies.
- Performs periodic asset discovery and audit tasks.
- Receives assets and ensure delivery to correct locations.
- Coordinates asset setup and teardown activities when requested.

Procurement Analyst

This role coordinates activities to accept deliveries of assets, performs initial validation of products received, identifies any discrepancies and physically tags assets for identification.

- Coordinates procurement activities when requested
- Ensures assets are shipped to correct locations
- Ensures assets are appropriately received and tagged
- Coordinates activities to dispose of assets when requested

License Administrator

This role is responsible for tracking, controlling and reporting on software licenses throughout the IT infrastructure. It ensures that licenses are being used per vendor licensing agreements and requirements.

- Handles procurement of software licenses when requested.
- Tracks all software licenses, their usage and owners.
- Administers licenses.
- Performs procurement tasks for software licenses.
- Assists with software license audit activities when requested.
- Ensures licenses are compliant with vendor usage specifications.
- Interfaces with Release Management activities and policies.

Transition Manager

This role has responsibility for identifying the transition strategy and detailed transition plans for migrating releases into production operations. It oversees all transition activities related to the deployment of a release.

- Identifies transition strategy for specific releases.
- Develops detailed transition plans.
- Obtains staffing resources to execute transitions.
- Oversees execution of transition plans and strategies.
- Reports status of transition efforts to Release Manager.
- Validates success of transition activities and plans.

Release Build Manager

This role is responsible for activities needed to configure, build, and package release for deployment. It also validates that packages have been adequately tested and that Known Errors are documented prior to production deployment.

- Manages and coordinates the development of release packages.
- Budgets and accounts for release deployment team activities and resources.
- Acts as the prime interface in terms of solution deployment planning and reporting.
- Makes a final recommendation regarding the decisions to release and deploy into production.
- Ensures all organizational policies and procedures are followed throughout deployment.
- Ensures release packages meet agreed customer and stakeholder requirements.
- Defines the requirements, processes and tools for release package deployment.
- Maintains and integrates lower level plans to establish overall integrated deployment plans, including planned vs. actual.
- Maintains and monitors progress on deployment related changes, issues, risks and deviations including tracking progress on actions and mitigation of risks.
- Provides management information on resource use, project/solution deployment progress, budgeted and actual spend.
- Coordinates release deployment activities across projects, suppliers and service teams where appropriate.

Deployment Analyst

This role supports the Transition Manager by building and customizing the releases to be implemented. It carries out detailed planning and execution activities to stage, distribute, and implement them.

- Assists with all aspects of the end-to-end deployment process.
- Updates (or requests updates to) the knowledge and configuration databases.
- Ensures coordination of build and test environment teams.
- Provides management information on deployment progress.
- Assists with deployment policy and planning.
- Assists with release package design, build and configuration.
- Assists with release package acceptance activities.
- Assists with service roll-out planning including method of deployment.
- Assists with release package testing to predefined acceptance criteria.
- Performs deployment communication, preparation and training activities.
- Audits hardware and software before and after the implementation of release packages.
- Installs new or upgraded hardware.
- Assists with release, distribution and the installation of packaged software when necessary.
- Establishes the final release configuration (e.g. knowledge, information, hardware, software and infrastructure).
- Builds the final release package.
- Establishes and reports outstanding Known Errors and workarounds.
- Provides input to the final implementation sign-off process for releases.
- Provides or assists with early release support activities as required.
- Ensures delivery of appropriate support documentation.

- Provides release acceptance for provision of initial support.
- Provides initial support in response to incidents and errors detected within a new or changed release package that has been implemented.
- Supports formal transition of releases to production operations.
- Provides early handholding support during release and deployment.

Knowledge Manager

This role plans, manages and controls all knowledge used to support provided services and the IT infrastructure. It ensures that needed knowledge is available to IT staff and business unit customers in accordance with security and functional needs to support their job functions.

- Establishes and manages the Service Knowledge Management System (SKMS).
- Undertakes the Knowledge Management role, ensuring compliance with documentation policies and processes.
- Performs knowledge identification, capture and maintenance activities.
- Identifies, controls, and stores any information deemed pertinent to the services provided.
- Oversees the controlled knowledge items to ensure accuracy and viability.
- Ensures all knowledge items are made accessible to those who need them in an efficient and effective manner.
- Monitors publicity recognizing the Service Knowledge management System as a central source of information.
- Acts as an adviser on Knowledge Management matters, including policy decisions on storage, value, worth etc.

Knowledge Architect

This role establishes the overall strategy of the Service Knowledge Management System (SKMS) and ensures a well-architected set of technical solutions and standards to support knowledge initiatives. This role also coordinates common technology related activities between all teams involved in the development and maintenance of knowledge.

- Ensures the knowledge architecture meets the strategic needs for delivering and supporting provided services.
- Coordinates knowledge technology product selections and tailoring.
- Ensures maximum integration of knowledge tools.
- Coordinates knowledge product implementation activities.
- Coordinates knowledge technology customization and integration activities.
- Coordinates technical resources to optimize use of knowledge technology solutions.
- Identifies ongoing support and maintenance for knowledge technologies chosen.
- Communicates chosen knowledge architectures and solutions to service delivery and support personnel.
- Interfaces to knowledge technology vendors as needed.

Knowledge Administrator

This role maintains the Service Knowledge Management System (SKMS) by responding to requests to update, add, backup, restore or delete knowledge assets in that system.

- Reviews, classifies, and reports on the integrated, meaningful information within the SKMS.
- Assists in the management and maintenance of knowledge repositories.
- Ensures SKMS is appropriately backed up.
- Restores and recovers knowledge assets when needed.

Knowledge Owner

This role is responsible for one or more areas of knowledge relevant to support, delivery and operation of the services provided by IT. It acts as a single point of contact for one or more knowledge areas and ensures that knowledge within those areas is accurate and maintained to current levels.

- Assists in identification and collection of knowledge assets.
- Reviews available knowledge on a periodic basis for accuracy and fit for use.
- Raises RFCs as needed to make knowledge changes.
- Acts as a single point of contact focus for knowledge areas responsible for.

Trainer

This role delivers and conducts training to IT and business staff.

- Uses best-in-class facilitation tools, techniques, and methods to conduct facilitated sessions.
- Demonstrates effective questioning and feedback techniques when delivering classroom instruction.
- Responds to participant need for clarification or feedback.
- Conducts training sessions.
- Provides training in conjunction with the rollout of new processes, procedures, systems, software, tools, and updated versions.
- Establishes and maintains credibility as a facilitator and subject-matter expert.
- Facilitates groups by applying knowledge of group dynamics, personality types and interpersonal interactions.

Training Administrator

This role designs and builds training curriculum and content. It may also coordinate development of training with outside vendors.

- Monitors attendance and attendee progress.
- Schedules training sessions and events.
- Sets up training events and classes.
- Prepares the delivery of training materials.
- Manages inventory of training materials.
- Procures outside training as needed.
- Handles communications between training providers and attendees to establish training classes and events.

Training Architect

This role performs training administration tasks such as scheduling of training, training lab setup, communication of training events, enrollment, tracking and monitoring compliance to training requirements. It may also coordinate delivery of training with outside vendors.

- Identifies the characteristics and needs of a specific target audience and tailors existing training to the needs of the audience.
- Updates/maintains training based upon training evaluation data and monitoring of the subject matter and the environment in which the associated skills and knowledge are applied.
- Identifies and evaluates appropriate instructional methods and motivational techniques to address different learning styles.
- Develops and provides job aids to support users of systems and tools as appropriate.
- Makes use of day-to-day tasks, underlying processes, and actual work environment contexts for instructional purposes.
- Delivers training solutions to meet the intended training objectives using prepared materials and activities.
- Works with SMEs to obtain instructional content and understand the environment in which the target audience works.
- Verifies training products are technically accurate, appropriately matched to the characteristics and needs of the target audience, instructionally sound, and aligned with user, regulatory, and legal requirements.
- Designs educational experiences that encourage critical reflection and alternative ways of thinking, behaving and working.
- Develops and implements training solutions using a variety of instructional strategies and delivery mechanisms.
- Creates supporting and supplemental training materials, including job aids.

- Develops training requirements.
- Evaluates training and training materials and identifies opportunities for improvement.
- Evaluates effectiveness and efficiency of training and performance interventions in achieving desired results, including content and delivery media.
- Evaluates and assesses instruction, including learner performance and delivery of instruction.
- Facilitates groups by applying knowledge of group dynamics, personality types and interpersonal interactions.

Organizational Change Leader

This role develops and leads the organizational change effort to alter business culture and behaviors towards alignment with new or changed processes and services. It monitors and oversees all stakeholders and carefully crafts and controls all key messages about IT services, projects and plans.

- Performs Stakeholder Management activities to identify Stakeholder concerns and issues with solutions being developed.
- Monitors stakeholder acceptance/rejection of solutions being developed.
- Crafts and controls key communications and messages about the implementation effort.
- Identifies opportunities to win acceptance of solutions being developed by those who are impacted.
- Identifies channels for communications and builds the overall communications plan.
- Develops a Resistance Management Plan to provide strategies for dealing with rejection or resistance to solutions being developed.
- Ensures appropriate levels of the organization are involved and demonstrating active commitment and leadership to the solutions being developed.
- Coaches senior management and other key personnel to help them "walk the talk" and demonstrate commitment to IT Service management solutions.

Technical Writer

This role documents process guides and work instructions in a manner that is easily understood by those executing the processes. It participates in the documentation of tool architectures and tool changes. It builds and publishes templates for presentations and key ITSM forms.

- Provides assistance in setting standards for how processes and procedures should be documented.
- Produces documentation for process guides, procedures, work instructions, technical documentation and other documented artifacts used in the IT infrastructure.
- Provides consulting guidance on how to best present documented information so it is quickly and easily understood.
- Identifies improvements for existing documentation.
- Designs and builds templates for key presentations and process work products
- Designs and builds templates for forms used as part of ITSM solutions.

Service Operation Roles

This set of roles support the ITSM Service Operation Lifecycle stage. Mappings of job functions addressed by these roles can be shown as follows:

Role	Job Function(s)
Incident Manager	Service Desk
Incident Analyst	Service Desk
Incident Auditor	Service Desk
Service Desk Manager	Service Desk
Service Desk Analyst	Service Desk
Service Desk Administrator	Service Desk
Call Agent	Service Desk
Service Desk Inf. Architect	Service Desk
Request Manager	Service Desk
Request Administrator	Service Desk
Request Fulfillment Owner	Can be any job function
Problem Manager	Service Desk
Problem Owner	Can be any job function
Monitoring Manager	Operations Control
Monitoring Architect	Technical Management
Security Administrator	Operations Control
Facilities Security Admin.	Facilities Management
Physical Site Manager	Facilities Management
Site Architect	Facilities Management
Site Contractor	Facilities Management
Site Technician	Facilities Management
Office Manager	Facilities Management
Network Operations Manager	Operations Control
Network Support Analyst	Operations Control
Network Technician	Technical Management
Network Architect	Technical Management
Network Administrator	Operations Control
Operations Support Manager	Operations Control

Role	Job Function(s)
Operations Support Analyst	Operations Control
Operations Architect	Technical Management
Scheduler	Operations Control
Storage Administrator	Operations Control
Technical Support Manager	Technical Management
Technical Support Analyst	Technical Management
Systems Administrator	Technical Management
Database Administrator	Technical Management

Incident Manager

This role is responsible for managing Incident staff and provides oversight for the Incident Management process. It ensures that incidents are handled and communicated in a timely manner and seeks to improve incident handling activities wherever possible.

- Drives the efficiency and effectiveness of the Incident Management process
- Produces Incident Management reports and information
- Manages the work of Incident support staff (first-and second-line)
- Monitors the effectiveness of Incident Management and making recommendations for improvement
- Develops and maintains Incident Management systems
- Establishes and executes ongoing audit activities within Incident Management

Incident Analyst

This role supports the Incident Management process by providing an operational single point of contact to manage incidents to resolution.

- Performs incident registration
- Routes incidents to support specialist groups when needed
- Analyzes for correct classification and provide initial support
- Provides ownership, monitoring, tracking and communication of incidents
- Provides resolution and recovery of Incidents not assigned to support specialist groups
- Closure of Incidents
- Monitoring the status and progress towards resolution of assigned Incidents
- Keeping Service Desk informed about incident progress
- Escalating incidents as necessary per established escalation policies

Incident Auditor

This role is responsible for auditing recorded incident records to ensure that incidents have been accurately classified and prioritized.

- Understanding current incident classification and prioritization criteria
- Conducting periodic and regular audits of incident tickets
- Identifying audit failures to Incident Management staff
- Producing audit reports on Incident quality

Service Desk Manager

This role is responsible for management, supervision, organization, and staff support capabilities of the Service Desk. It ensures that the Service Desk is operating with the Incident Management and Request Fulfillment processes. It reports on quality of call management activities.

- Provide leadership to Service Desk staff to develop and meet Service Desk goals and strategies.
- Provide call management and support services in line with Service Level targets.
- Counsel and Coach Service Desk staff.
- Perform periodic Service Desk staff performance reviews.
- Review and analyze all Service Desk and Incident Management reports to proactively seek improvements.
- Develop and communicate Service Desk staff training and skills maintenance plans.
- Work with key Service Desk Customers to address customer related issues.
- Oversee actions to obtain feedback from customers on IT service quality
- Administer and manage staffing levels in line with IT service needs.
- Ensure all incidents are being addressed and appropriately escalated to support staff.
- Manage all Service Desk staff recruiting activities.
- Maintain Service Desk staff morale and strive for low staff turnover rates as much as possible.
- Represent Service Desk functions on Change Management CAB meetings

Service Desk Analyst

This role provides leadership and mentoring for Service Desk Call Agents to resolve Service Desk issues and maintain customer satisfaction at high levels.

- Mentor Service Desk Call Agents.
- Recommend resolutions to technical problems reported by customers not easily resolved by Service Desk Call Agents.
- Consult with peers on technical issues pertaining to all systems and applications.
- Audit Incident Management Database to ensure all Incidents are being logged and categorized accurately.
- Maintain high level of customer satisfaction.
- Act as a focal point for disseminating communications about services and policies among Service Desk staff.
- Provide appropriate communications and turnover for long running incidents as needed.
- Assist with actions to improve Service Desk services as requested.
- Ensure all incidents are logged in the Incident Management Database.
- Act as liaison between customer and user groups to support them in jointly resolving incidents.
- Identify needs related to training, documentation, and technical issues.
- Train new Service Desk staff members in Service Desk operating procedures
- Assist in preparation of staff schedules to ensure Service Desk is staffed appropriately to cover all hours of operation.

Service Desk Administrator

This role provides administrative support for Service Desk activities at the direction of the Service Desk Manager.

- Administer Call Management systems.
- Collect and gather Call Management data for reporting purposes.
- Assist with Service Desk incident communications as needed.
- Maintain Service Desk escalation and contact lists.
- Administer Service Desk Knowledge Bases as needed.
- Administer changes to incident handling procedures as directed by the Service Desk Manager.
- Prepare Service Desk training guides and materials with content provided by the Service Desk Manager.
- Assist with other tasks as directed by the Service Desk Manager.

Call Agent

This role handles calls for requests and incidents from customers and end-users to maintain high levels of satisfaction with IT services.

- Maintain end to end responsibility for customer calls providing timely, reliable and courteous service.
- Provides customer service and first level technical resolution for operational and service-related incidents.
- Resolve or escalate incidents and requests in line with established Service Level targets.
- Support Security Management activities by exercising constant vigilance for possible security implications during customer interactions.
- Provide feedback of intelligence gained through customer interactions.
- Maintain appropriate level of skills to handle incidents and requests in line with established service levels.
- Log and record all reported incidents into the Incident Management Database.
- Respond to all customer requests with accurate and appropriate information.
- Identify improvements to Service Desk services and operation on an ongoing basis.

Service Desk Infrastructure Architect

This role designs and maintains the Service Desk infrastructure that is used to support Call Management and incident handling activities.

- Design and oversee development of Call Management infrastructure.
- Design and oversee development of monitoring infrastructure for the Service Desk.
- Recommend changes and improvements to Service Desk support tools.
- Liaison with 3rd party vendors and other IT organizations that underpin Service Desk functions.
- Resolve incidents related to Service Desk infrastructure errors.

Request Manager

This role is responsible for managing request handler staff and execution of the Request Fulfillment process.

- Provide leadership to Request Management staff to develop and meet Request Management goals and strategies.
- Handle staff, customer and management concerns, problems and inquiries.
- Provide Request Management services in line with Service Level targets.
- Counsel and coach Request Management staff.
- Perform periodic Request Management staff performance reviews.
- Review and analyze all Request Management reports to proactively seek improvements.
- Develop and communicate Request Management staff training and skills maintenance plans.
- Oversee actions to obtain feedback from customers on quality of Request handling services.
- Administer and manage staffing levels in line with IT service needs.
- Ensure all requests are being addressed, appropriately escalated to support staff and fulfilled on a timely basis.
- Manage all Request Management staff recruiting activities.
- Maintain Request Management staff morale and strive for low staff turnover rates as much as possible.
- Represent Request Management functions on Change Advisory Board (CAB) meetings.
- Reviews the initial prioritization of requests to determine accuracy & consistency.

Request Administrator

This role handles requests from business users to maintain high levels of satisfaction with IT services. It oversees, manages and coordinates all activities to respond to a request and serves as a single point of contact for a request until a resolution has been achieved.

- Provides single point of contact and end to end responsibility to ensure submitted requests have been fulfilled.
- Provides initial triage of requests to determine which IT unit will fulfill them.
- Communicates requests to fulfillment staff.
- Resolve or escalate requests in line with established Service Level targets.
- Provide feedback of intelligence gained through customer interactions.
- Log and record all submitted requests.

Request Fulfillment Owner

This role provides a single point of contact into IT for the Request Handler to fulfill requests that have been submitted

- Provides single point of contact into IT to for fulfilling requests.
- Coordinates actions and activities to ensure that requests are fulfilled.
- May assign staff to work on requests.
- Communicates request progress to Request Management staff.
- Communicates and coordinates request fulfillment activities with requestors as needed.

Problem Manager

This role is responsible for reviewing problem trends and proactively taking actions to identify problems and reduce recurring incidents by removing errors from the infrastructure.

- Produces Problem Management reports and management information
- Manages Problem support staff
- Allocates resources for resolving problems
- Reviews the efficiency and effectiveness of proactive Problem Management activities.
- Identifies trends and potential Problem sources (by reviewing Incident and Problem analyses)
- Prevents the replication of Problems across multiple systems

Problem Owner

This role provides a single point of contact for one or more problems and is responsible for ownership and coordination of actions to analyze for root cause, identify Known Errors and coordinate activities to remove errors from the infrastructure.

- Reviews Incident data to analyze assigned problems
- Investigates assigned problems through to resolution or error identification
- Coordinates actions of others as necessary to assist with analysis and resolution actions for problems and Known Errors
- Raises RFCs to clear errors
- Monitors progress on the resolution of Known Errors and advises Incident Management staff on the best available Work-Around for Incidents related to unresolved Problems/Known Errors
- Assists with the handling of major Incidents and identifying the root causes

Monitoring Manager

This role manages day-to-day event management functions to provide automated and timely notification of events to IT and business unit staff.

- Provides single point of ownership for effective provision of event monitoring and management services to customers.
- Oversees monitoring activities and services provided to support teams.
- Oversees design tasks related to development of new monitors and alerts.
- Designs and develops filtering criteria for events.
- Designs and develops escalation criteria and routing paths for events.
- Manages and develops event management support staff.
- Oversees recruitment of event management support staff.
- Assigns event management support staff to projects and initiatives as needed.
- Resolves event management issues and design decisions that have been escalated.

Monitoring Architect

This role designs and executes automated capabilities to capture events (IT alarms, alerts and notifications) in the IT infrastructure and forward them to appropriate personnel or systems for further action.

- Collects and coordinates IT infrastructure events and event triggers to be acted upon
- Prioritizes, escalates and forwards events to personnel or systems to be acted upon
- Provides filtering and event correlation mechanisms to reduce event noise and ensure events are forwarded to their root sources for action
- Broadcasts and displays events to other parties and systems that may need to be aware of them upon request
- Provides event logging and historical repositories to aid in the investigation of incidents, problems and overall service quality
- Plans, designs, builds, tests, implements event management toolsets and manage automated responses to events to prevent or reduce service outages and unplanned labor
- Maintains event tables and rules in accordance with changes to the IT infrastructure
- Processes requests for event monitoring and handling in accordance with a consistent request fulfillment process
- Designs, builds, implements, tests and maintains console solutions to consolidate views of service events or highlight them for support staff
- Provides design and operational consulting to operational centers as needed

Security Administrator

This role is responsible for managing user IDs, passwords and access control lists that allows IT and business staff access to services and service assets.

- Administers security tools.
- Assists with communications of security-related policies and information.
- Monitors for security violations and incidents.
- Logs all security incidents.
- Supports Incident and Problem Management activities from a security perspective.
- Assists in the resolution of security incidents and problems.
- Assists with security audits when requested.
- Assists with security testing when requested

Facilities Security Administrator

This role provides ongoing management and maintenance of the physical security access infrastructure used to control access to IT processing facilities and data centers.

- Oversees physical site security (badges, cameras, key locks, etc.) management activities.
- Oversees design tasks related to execution of new physical site security policies and procedures.
- Manages and recruits and schedules physical security site administration staff.
- Escalates security violations and issues to executive staff when necessary.
- Provides assistance to forensic activities as needed to resolve security issues.
- Manages and controls physical access to secure areas per business unit specifications.
- Provides sponsorship of facility clearances.
- Manages the safeguarding and handling of classified information at the physical site
- Responsible for orchestrating security inspections, security training, security briefings/debriefings, and required reporting.
- Conducts periodic security self-inspections as needed to maintain facility security integrity.
- Serves as liaison between the business and other organizational security officers.

Physical Site Manager

This role provides ongoing management and maintenance of Data Center and other processing physical site infrastructures used to house IT hardware, supplies and people.

- Provides single point of ownership for monitoring of physical site events
- Oversees monitoring activities and services provided to support teams.
- Oversees design tasks related to development of physical monitoring system alerts.
- Designs and develops filtering criteria for facility monitoring system events.
- Designs and develops escalation criteria and routing paths for events.
- Manages and develops facility management support staff.
- Oversees recruitment of facility management support staff.
- Assigns facility management support staff to projects and initiatives as needed.
- Resolves escalated facility event management issues and design decisions.
- Monitors use of physical site premises power and utilities to identify upgrades or downgrades to accommodate changes in equipment capacity, local building codes, zoning, people or general physical infrastructure.

Site Architect

This role is responsible for designing all infrastructure physical site assets such as building core and shell, floor space, cabling, fire protection systems, mechanical, electrical, lighting, raised floor, desks, cubicles, console shelving and mounting.

- Provides consulting and expertise for physical site infrastructure assets such as floor space, mechanical, electrical, ventilation, cabling, lighting, raised floor, desks, cubicles, console shelving and mounting
- Provides consulting and expertise for local building and operating codes and zoning (safety) ordinances
- Oversees development of site blueprints and floor layouts
- Assists in physical site infrastructure requirements design considering equipment power and weight loads as well as needs for future floor space and environmental capacity
- Assists in designs for rehabilitation of existing building infrastructure to ready it for data center use
- Oversees site contractor activities to ensure site is constructed and/or modified per desired architectural specifications

Site Contractor

Procures and coordinates all external vendor activities to implement the physical building infrastructure, floor space, cabling, fire protection systems, mechanical, electrical, lighting, raised floor, desks, cubicles, console shelving and mounting.

- Procures and coordinates external vendor tasks to provide physical site infrastructure, mechanical, electrical, ventilation, cabling, lighting, raised floor, desks, cubicles, console shelving and mounting activities.
- Ensures site installation activities meet local standards for building and operating codes and zoning (safety) ordinances.
- Manages and coordinates external vendor installation schedules and scheduling dependencies.
- Resolves installation issues and conflicts if they arise.
- Oversees changes to physical site infrastructure to ensure existing services are not adversely impacted by physical site construction activities.
- Reports on infrastructure build progress and issues.

Site Technician

Manages and maintains all infrastructure physical site assets such as building core and shell, floor space, cabling, fire protection systems, mechanical, electrical, lighting, raised floor, desks, cubicles, console shelving and mounting.

- Provides consulting and expertise for physical site infrastructure assets such as floor space, mechanical, electrical, ventilation, cabling, lighting, raised floor, desks, cubicles, console shelving and mounting.
- Manages and maintains physical site locations to provide clean operating environment free from litter, dust and pollutants.
- Manages, tests, and maintains fire protection systems.
- Monitors use of physical site premises power and utilities to identify upgrades or downgrades to accommodate changes in equipment capacity, local building codes, zoning, people or general physical infrastructure.
- Oversees changes to physical site infrastructure to ensure existing services are not adversely impacted by physical site construction activities.
- Oversees repairs to physical site infrastructure components done by 3rd parties and validate that repairs meet expected benefits.
- Ensures proper labeling of equipment, cables and telecommunication lines is in place and adequately maintained.

Office Manager

This role oversees day-to-day activities to run and support office infrastructure for the IT Data Centers as well as other support locations.

- Recruits office support staff such as secretaries, administrators and receptionists.
- Designs and plans out use of office spaces.
- Readies office space for staff on a timely basis.
- Designs and plans staff permits such as for use of parking facilities and loading dock.
- Procures and coordinates office external services such as garbage pickup, fire department and utilities.
- Coordinates activities to put office infrastructure into place including cubicles, telephones, FAX, Printers and lounge facilities.
- Oversees day-to-day activities of reception and secretarial staff.
- Maintains inventory of office supplies
- Manages and schedules access to physical site conference and meeting rooms.
- Manages and maintains conference and meeting room supplies, audio, video and teleconferencing equipment.
- Coordinates visits to the facility in cooperation with the Facility Security Administrator.

Network Operations Manager

This role manages day-to-day network infrastructure operations to underpin delivery of systems and services in order to meet or exceed agreed services levels.

- Provides single point of ownership for effective provision of networking services to customers.
- Oversees network operational activities and services for one or more operational delivery centers.
- Proactively identifies and implements service improvements in network operational delivery centers.
- Approves acceptance into production of new network configurations, network changes and services.
- Ensures service continuity plans are compatible with network operational delivery center operations and that plans are tested on a regular basis.
- Oversees the NOC (Network Operations Center)
- Manages and develops all network operational delivery center staff.
- Oversees recruitment of network operational staff.
- Manages network operational delivery center budget.
- Approves procurement of new networking CIs that will reside in the operational delivery center.
- Provides service reports on network operational delivery center (NOC) performance and quality.

Network Support Analyst

This role is responsible for performing all network operational processes and procedures, ensuring that all networking services and infrastructure meet their operational targets.

- Operates and implements all networking operational infrastructure and procedures.
- Participates in incident and problem support activities when requested.
- Investigates, diagnoses, and takes prescribed actions on all networking events, alarms and incidents.
- Monitors all networks to ensure service quality is being delivered on a daily basis.
- Maintains operational logs and journals on all networking events, warnings, alerts and alarms, recording and classifying all messages.
- Maintains all networking data collection procedures, mechanisms and tools.
- Maintains all networking operational documentation, processes, management and diagnostic tools and spares, ensuring that spares are maintained at the agreed levels.
- Ensures that all routine network maintenance tasks are completed.
- Ensures that all networking infrastructure equipment is maintained according to policies and recommendations and performs regular checks on environmental equipment and conditions.

Network Technician

This role manages, supports, and maintains the networking and telecommunications infrastructure.

- Installs, configures and tests networking infrastructure components and connectivity
- Decommissions networking infrastructure components upon request
- Maintains networking components in compliance with supporting vendor requirements
- Provides consulting services and support for release package testing, installation, deployment and operation
- Provides troubleshooting and technical support services for the networking infrastructure, IP addresses and naming services
- Provides consulting and support services to identify network operational, monitoring and reporting requirements
- Coordinates, schedules, implements and tests network tuning activities
- Implements and tests network security schemas per requested requirements

Network Architect

This role is responsible for designing the networking and telecommunications infrastructure used to underpin IT services.

- Designs networking infrastructure components and connectivity
- Identifies network traffic load, impact and capacity requirements needed to support services
- Identifies physical facility requirements needed to operate networking components (i.e. floor space, equipment clearance, electrical, cooling, cabling, weight load)
- Provides networking requirements to support network procurement activities.
- Provides consulting services and support for release package testing, installation, deployment and operations.
- Provides troubleshooting and technical support services for the networking infrastructure, IP addresses and naming services
- Provides consulting and support services to identify network operational, monitoring and reporting requirements
- Defines network security schemas per requested requirements.

Network Administrator

This role performs network hardware and software administrative tasks used to support the networking infrastructure.

- Maintains information about network topology and configuration items and ensure accuracy and availability to others.
- Provides network configuration information for release package testing, installation, deployment and operations.
- Labels network components with asset tags and track component locations, serial numbers, IP/MAC addresses and owners.
- Coordinates and schedules server repair services with 3rd party vendors and validate that expected repairs and network patches achieved expected benefits.
- Manages and maintains definitive hardware stores for networking spare parts and equipment.

Operations Support Manager

This role manages day-to-day infrastructure operations to provide delivery of systems and services in order to meet or exceed agreed services levels.

- Provides single point of ownership for effective provisioning of systems and services.
- Oversees operational activities and services for one or more operational delivery centers.
- Proactively identifies and implement service improvements in operational delivery centers.
- Approves acceptance into production of new systems and services.
- Ensures recovery and continuity plans are compatible with operational delivery center operations and that plans are tested on a regular basis.
- Represents interests of operational delivery centers managed for all service initiatives.
- Manages and develops all operational delivery center staff.
- Oversees recruitment of operational staff.
- Manages operational delivery center budgets.
- Approves procurement of new CIs that will reside in the operational delivery center.
- Provides service reports on operational delivery center performance and quality.

Operations Support Analyst

This role is responsible for performing all operational processes and procedures, ensuring that all services and infrastructure meet their operational targets.

- Operates and implements all operational infrastructure and procedures.
- Participates in incident and problem support activities when requested.
- Investigates, diagnoses, and takes prescribed actions on all operational events, alarms and incidents.
- Monitors all operations and services to ensure service quality is being delivered on a daily basis.
- Maintains operational logs and journals on all events, warnings, alerts and alarms, recording and classifying all messages.
- Maintains all operational data collection procedures, mechanisms and tools.
- Maintains all operational documentation, processes, management and diagnostic tools and spares, ensuring that spares are maintained at the agreed levels.
- Ensures that all routine maintenance tasks are completed on all operational infrastructures.
- Ensures that all infrastructure equipment is maintained according to policies and recommendations and performs regular checks on environmental equipment and conditions.

Operations Architect

This role is responsible for the overall coordination and design of tools and technologies used to underpin operational activities.

- Designs secure and resilient operational infrastructures that underpin operational solutions and activities.
- Maintains all operational design, architectural, policy and specification documentation.
- Identifies opportunities for automation for manual and redundant operational tasks.
- Recommends solutions to continually improve operations on an ongoing basis.
- Provides advice and guidance to analysts, planners, designers and developers on all aspects of operational designs and technologies.
- Interfaces with designers and planners from external suppliers and Service Providers, ensuring all external operational services are designed to meet their agreed service levels and targets.
- Reviews and contributes to the design, development and production of new services, Service Level Agreements (SLAs), Operation Level Agreements (OLAs) and Underpinning Contracts (UCs), covering equipment and services.
- Maintains a good technical knowledge of all installed operational product capabilities and the technical frameworks in which they operate.

Scheduler

This role is responsible for the management and control of all aspects of the scheduling, monitoring and control of operational workloads.

- Prepares and maintains day-to-day operational workload schedules in line with scheduling guidelines.
- Ensures that operational workloads are run according to their defined schedules.
- Processes ad hoc workload requests when requested and approved.
- Administers scheduling tools and infrastructure.
- Develops and maintains all necessary operational scheduling documentation.
- Produces workload scheduling reports that report results of schedules and job runs in a timely fashion.

Storage Administrator

This role is responsible for the management and control of storage media, backup and recovery schedules, testing, storage planning, allocation, monitoring and decommissioning of storage devices.

- Interfaces with Availability, Capacity, Security and IT Service Continuity Management to ensure that all requirements are met by current backup and recovery policies.
- Develops and manages a Data Retention Policy that is compliant with legal and regulatory requirements.
- Implements and administers backup and recovery packages and tools.
- Procures magnetic tapes, diskettes, cartridges, paper, microfiche and all other media and devices when required.
- Manages and maintains media pick lists and vaulting mechanisms.
- Establishes and maintains a clear physical identification system for media for easy identification.
- Monitors backup jobs and schedules to ensure these take place without error.

Technical Support Manager

This role manages day-to-day technical support functions to provide delivery of expertise and support to infrastructure teams in order to meet or exceed agreed services levels.

- Provides single point of ownership for effective provision of technical support services to customers and stakeholders.
- Oversees technical support activities and services provided to support teams.
- Oversees design tasks related to development of new systems and services.
- Manages and develop all technical support staff.
- Oversees recruitment of technical support staff.
- Assigns technical support staff to projects and initiatives as needed.
- Manages technical support budget.
- Resolves technical support issues and design decisions that have been escalated.

Technical Support Analyst

Supports IT infrastructure solutions with technical skills expertise to maintain steady state service operations or design tasks for new services.

- Analyzes infrastructure technical issues.
- Resolves infrastructure incidents and problems when they occur.
- Maintains ownership of problem diagnosis, resolution and escalation for all received problems and issues.
- Participates in service design and transition technical activities for new or changed services when needed.
- Applies third party maintenance to infrastructure components.
- Develops and maintains infrastructure documentation and procedures.
- Provides technical support and assistance with feasibility studies.

Systems Administrator

This role administers and maintains infrastructure devices such as servers, hosts and networking devices to ensure proper operation and availability.

- Controls and administers hardware and operating software configurations.
- Monitors devices for proper operation and performance.
- Applies vendor provided maintenance to devices.
- Detects, diagnoses, isolates and corrects device operational failures.
- Supports Incident and Problem Management activities with device expertise and troubleshooting.
- Maintains awareness of new technologies that might enhance device operation, capacity and performance.

Database Administrator

This role is responsible for the management and control of databases in the infrastructure.

- Performs database physical and logical design tasks to meet the objectives new IT solutions and services.
- Implements database logging and recovery operations.
- Works with Capacity Management to perform database sizing and support capacity workload and forecasting estimates for databases.
- Implements database backup and recovery procedures.
- Monitors databases for adequate performance and capacity.
- Assists IT developers with database architecture and access control policies.
- Identifies appropriate database solutions and products.

Continual Service Improvement Roles

This set of roles support the ITSM Service Operation Lifecycle stage.

Role	Job Function(s)
Service Manager	Service Management
Process Owner	Process Management
Service Owner (Product Manager)	Service Management
Business Relationship Manager	Service Management
Reporting Architect	Service Reporting
Reporting Administrator	Reporting Administrator
Quality Assurance Analyst	Quality Assurance

Service Manager

This role oversees the entire IT Service Management (ITSM) operation to ensure that quality service management solutions are developed and deployed to meet agreed business objectives.

- Manage ITSM staff and budget.
- Enable and champion an IT service culture.
- Develop, implement and maintain ITSM-based management processes and controls to ensure service quality is maintained to meet business objectives.
- Champion and promote service improvements on an ongoing basis to continually improve quality and customer satisfaction with IT services.
- Maintain day to day responsibility for the ownership and resolution (including any referral or escalation as may be necessary) of Service Management issues which arise in connection with ITSM Services.
- Review service metrics (KPIs—Key Performance Indicators) that identify the success of the services being utilized to recommend and coordinate implementation of changes to ITSM services to improve metrics.
- Work to ensure continuous alignment of the services with the customers' needs, i.e. changing work patterns, workloads, revised aims and objectives.
- Co-ordinate inter-process changes with ITSM process owners.
- Ensure alignment of ITSM solutions to the corporate and IT strategy.

Process Owner

Owns one or more processes and is responsible for process quality and coordinating process with other processes in the organization.

- Responsible for the overall process objectives.
- Provides direction to staff using the process.
- Monitors process maturity and progress.
- Coordinates design decisions and activities with other Process Owners.
- Assists in development of project work plans, schedules and staffing requirements from a process perspective.
- Communicates as required to executive management and Program Office.
- Ensures that process implementation and design requirements are adequately identified and that process solution issues are being addressed.
- Identifies process and solution requirements to Technical Architecture Teams.
- Ensures people are using the process effectively.
- Coaches and teaches others about process concepts and solutions.
- Communicates the process throughout all support teams.
- Provides overall leadership and management from a process perspective.

Service Owner (Product Manager)

Owns one or more services and is responsible for end to end service quality across the organization.

- Provides end to end responsibility for the service.
- Provides direction to staff using the service.
- Defines, plans, purchases and manages elements of the service and its performance.
- Monitors overall service delivery quality.
- Initiates service improvement actions on a continual basis.
- Responds to service deficiencies, problems and issues.
- Assists in development of project work plans, schedules and staffing requirements from a service perspective.
- Communicates as required to executive management and Program Office.
- Ensures that service implementation and design requirements are adequately identified and that service solution issues are being addressed.
- Identifies service requirements to Technical Architecture Teams.
- Monitors customer satisfaction with the service.

Business Relationship Manager

This role provides a single point of contact to one or more business units for all IT services. It reviews quality of services delivered to those units, addresses customer issues and communicates changes in customer needs back to IT.

- Acts as a single point of contact for one or more business units for all IT services
- Communicate service needs from the business units represented back to IT
- Escalates business unit service issues to the IT Program Office
- Documents requests on behalf of business units for complex and unique IT services and requirements not in the IT Service Catalog
- Communicates service status on service issues being addressed by IT back to the business unit
- Assists in service negotiation efforts between IT and the business unit
- Assists in coordinating service improvement projects, transition projects or other related efforts specific to one or more business units
- Reviews quality of services rendered to the business units being represented

Reporting Architect

This role designs and builds the reporting infrastructure used to collect and store metrics and generate service, process and resource reports used to make decisions on improvement actions.

- Works with other IT service management staff to identify needed metrics and key performance indicators and reports
- Identifies sources of metrics data and needed processing and calculations to produce reports
- Installs and customizes service management reporting tools
- Builds and documents data gathering and report production processes
- Identifies roles and responsibilities for reporting
- Builds and documents report repository infrastructures
- Responds to ad-hoc requests for production of new reports or report changes
- Identifies and implements distribution mechanism for distributing reports

Reporting Administrator

This role provides assistance in collection of metrics, key performance indicators and other data as well as generation of service, process and resource reports.

- Executes processes and procedures to collect metrics data and produce service improvement reports
- Ensures reports are accurately distributed to their recipients
- Identifies reporting concerns and issues as needed to the Reporting Architect
- Maintains report distribution lists in accordance with Change Management process
- Communicates user desired report changes to the Reporting Architect as needed
- Maintains report repository infrastructures
- Responds to ad-hoc requests to reissue reports

Quality Assurance Analyst

This role provides assistance in validating that IT services meet their stated requirements for utility (fit for purpose) and warranty (fit for use). It also ensures that IT services are built and delivered using desired means and methods for service design, transition and operation.

- Responsible for building and maintaining the Quality Control Plan
- Reports on attainment against quality plans and targets
- Supports, facilitates, and coordinates activities to independently verify that desired means and methods for service design, transition and operation are being used
- Ensures that services are compliant with external regulatory, security, legal and industry requirements and standards
- Ensures services are meeting stated utility requirements
- Ensures services are meeting stated warranty requirements
- Identifies and reports quality deficiencies to the Service Manager

Chapter

6

Process Architecture

Process Architecture Overview

This chapter summarizes the ITSM processes as they are presented in current ITSM literature. Since much industry literature exists that already describe ITSM processes, they are only summarized here.

The download website with this book provides a complete set of process activities and design considerations for putting the process architecture together. These are meant to be reference materials, however, and not how-to guides.

This chapter also includes a Process Meta-Architecture description and set of building blocks. These are used to describe what elements need to be in place for any process for it to be successfully defined, documented, and operated.

Included also, are some suggested techniques for aligning your current infrastructure documentation such as work instructions and procedures with your ITSM processes.

A set of building blocks for the process architecture are illustrated below. These align closely to the ITSM Service Lifecycle.

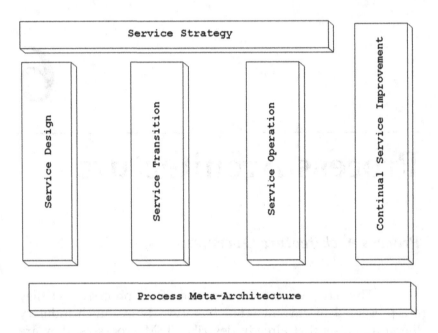

Figure 19: Process Architecture Model

Each building block shown above can be broken down into one or more ITSM processes. The following sections of this chapter provide a high level description of each building block shown above.

Service Strategy Processes

This grouping of processes support the identification of IT services to be provided, strategies for their delivery and their value equation to the business. The key processes are:

Service Portfolio Management

Creates and manages the Portfolio of IT Services through an ongoing set of portfolio management disciplines. The portfolio covers all IT services from their inception to retirement and identifies priorities and investments that make up the IT strategy.

IT Financial Management

Manages the IT budget, accounts for service costs and recovers those costs via charging the business for the IT services delivered.

Demand Management

This process identifies patterns of business activity that impact the usage of IT services. It also manages efforts to control that demand to better balance use of service resources.

Strategy Development

This process is used to develop IT strategies that translate into value creation through services. Key activities include identifying markets and customers, developing the service portfolio, developing the service assets and preparing to establish those services.

Business Relationship Management

This process is used to ensure high levels of customer satisfaction by providing a direct interface between the IT organization and the customer. Key activities include reviewing service quality delivered with customers, communicating and coordinating service issues, informing customers about available services to match their needs and identifying changing customer needs back to the IT organization.

Service Design Processes

This grouping of processes supports the design activities for new or changed IT services. The key processes are:

Design Coordination

Provides a single point of coordination and control for all IT service design activities and processes. Key activities include coordinating designs across multiple process, technology and organizational domains, enforcing design standards and methods, and overseeing production of design packages and specifications.

Service Catalogue Management

Creates and manages the IT Service Catalogue that describes all services available to users. Key activities include managing the catalog content, providing channels to communicate services and ensuring the information communicated about services is accurate and reliable.

Service Level Management

Identifies and manages the overall quality of services being delivered through an ongoing cycle of agreeing, negotiating, monitoring and reporting on service levels. Puts service agreements into place and initiates service improvement projects when deficiencies are found.

Capacity Management

This process ensures that enough resources are in place at the right time and at acceptable costs to support the anticipated demand levels for services.

Availability Management

This process ensures that IT services are available when needed by addressing infrastructure needs for availability and minimizing risks for service outages.

IT Service Continuity Management

This process ensures that IT services can be continued in the event of a major business interruption. It works to put continuity plans into place, maintain awareness of the plan and test plans on a periodic basis.

Information Security Management

This process ensures that IT services are designed to be used safely and reliably minimizing security risks. It also ensures that services are designed in compliance with required security policies and controls.

Supplier Management

This process ensures that external vendor suppliers to IT are appropriately managed to ensure they meet their service obligations.

Service Transition Processes

This grouping of processes supports the activities to build . and transition service designs and changes into a production state. The key processes are:

Transition Planning and Support

Creates and manages plans and strategies to transition changes and releases into production. Strategies chosen may be based on complexity, size or impact to minimize disruption to existing services.

Change Management

Manages and coordinates all changes made to the IT infrastructure to minimize risk and impact of those changes on service quality and availability.

Service Asset and Configuration Management

This process provides and maintains a logical model of the IT infrastructure ensuring that information on IT service assets and their relationships is accurate and available to IT support and delivery staff.

Release and Deployment Management

This process takes a holistic view of IT changes ensuring that all infrastructure components for those changes that need to be released together are identified, packaged, and deployed.

Service Validation and Testing

This process ensures that IT services are adequately tested to ensure they meet functional and operational requirements to deliver expected value.

Change Evaluation

This process manages and controls activities that provide a consistent and standardized means for determining the performance of a service change to make sure that business expectations are managed and any deviations between the predicted performance and actual performance are known and managed.

Knowledge Management

This process manages the activities to gather, analyze, store and share knowledge information needed by others in the support and delivery organization to effectively perform their functions.

Service Operation Processes

This grouping of processes supports the activities to manage and deliver IT services on a day-to-day basis. The key processes are:

Event Management

This process monitors all events that occur in the IT infrastructure to detect normal operation or to escalate exception conditions.

Incident Management

This process coordinates activities to restore normal service operation as quickly as possible when incidents occur.

Request Fulfillment

This process provides a channel for users to request and receive standard IT services that are pre-approved and coordinates activities to respond to those requests.

Problem Management

This process coordinates activities to identify root causes for incidents and proactively coordinate actions to remove those causes from the infrastructure to prevent recurrence of incidents.

Access Management

This process ensures only authorized users can access a service or group of services by administering security and availability policies and actions in the IT infrastructure.

Continual Service Improvement Processes

This grouping of processes proactively seeks to measure services, processes and resources and identify improvement actions on an ongoing basis.

7-Step Improvement Process

This process coordinates activities to identify measurements, gather them, analyze, process and present them to management for consideration of ongoing improvement actions.

Service Reporting

This process coordinates activities to create and publish service reports that identify the security, reliability, availability and overall quality of the services being delivered to the business.

Service Measurement

This process coordinates activities to define the quality measurement framework and service metrics that identify the security, reliability, availability and overall quality of the services being delivered to the business.

Continual Service Improvement

This process provides a continual repeatable approach for implementing service improvement projects by first identifying the improvement vision, assessing current state, targeting future state, planning the improvement, measuring the results and keeping the momentum going.

Return on Investment

This process provides a continual repeatable approach for identifying costs and benefits of improvement actions and justifying them to the business.

Process Meta-Architecture

The purpose of the process meta-architecture is to identify all the key sub-components that describe how processes are assembled and put together. Every process should include all the sub-components else, there is a significant risk that the process will not be effective or used by support staff.

The process meta-architecture can best be shown as follows:

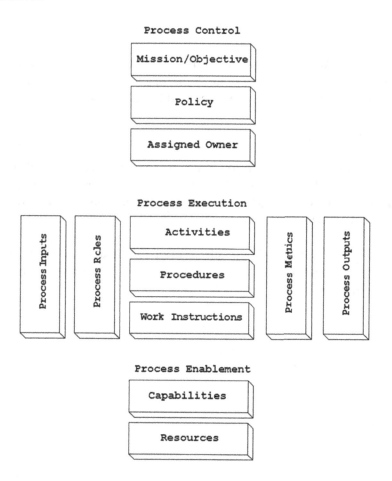

Figure 20: Process Meta-Architecture

Process Control

These sub-components describe the overall guiding controls for the process. They are:

Mission/Objective

This identifies the key business reasons for having the process in the first place. It describes the value that the process provides to stakeholders and customers.

Policy

Identifies the general business rules and principles by which the process will operate. Policies are critical to ensure that process activities and procedures will be carried out.

Assigned Owner

Every process must have an assigned owner who is accountable for the overall quality of the process, ensuring it is providing value and ensuring it is appropriately used by IT support and delivery staff.

Without the above sub-components, staff may view use of the process as wasteful, not important and may bypass it altogether.

Process Execution

These sub-components describe the activities, inputs, outputs and workflows for a given process. They also identify how the process will be measured and what organizational roles and responsibilities exist for carrying out the process activities. They are:

Activities

These identify the high level tasks that need to be accomplished for any given process. They describe what needs to be done at a high level.

Procedures

These break down each high level activity into sequences of sub-activities that need to be performed. It is common to see procedures described in many ways such as flowcharts or step diagrams. The goal of any procedure is to identify the workflow for how a process activity will get done.

Work Instructions

These provide the lowest granular level of instructions for identifying how any step in a procedure is carried out. Typically, these tend to be detailed steps to be taken to accomplish a procedural task or activity.

Inputs

These identify the triggers and prerequisites needed to set a process in motion. Examples of inputs may be infrastructure events that have been captured or a form that has been submitted.

Outputs

These identify the work products and artifacts that are produced because of the process execution activities. Examples of these could be forms or reports.

Process Roles

These identify specific groupings of related activities and tasks associated with any given process. Activities are grouped under a logical role title and responsibilities are assigned to each activity or task associated with the role. Assignments are then used to indicate whether the role is accountable, responsible, consulted or informed for each task that is part of that role. Once defined, roles are then assigned to job functions within your IT and business organization.

Process Metrics

These identify operational and key performance indicator measurements that are used to indicate how well the process is performing.

Without the above sub-components, there may be confusion over how a process should work, confusion over responsibilities and no evidence that the process is being used or meeting its mission/objective.

Process Enablement

These sub-components describe the service assets that are assigned to any given process. They are:

Resources

These identify hard assets such as people, hardware, software, management tools, facilities and financial capital.

Capabilities

These identify soft assets such as available skills, service culture, knowledge, management culture and process capabilities.

Without the above sub-components, processes may only exist as documents that are never executed or the process itself may execute poorly due to lack of resources.

Pulling Process Architecture Together

Process information is everywhere. It may exist in separate documents, stored in someone's file drawer or be inherently described in toolsets being employed. When putting process architecture together for the first time, you may find much of this information scattered all over the infrastructure in bits and pieces.

Each of these disparate items can be tied together via use of a common process architecture by:

1. First identifying the processes and activities that are in your architecture;
2. Identifying the procedures that describe each of those activities;
3. Linking your disparate items to the appropriate step that they relate to in your procedures;
4. Locking disparate items down that are selected via Configuration and Change Management processes.

This approach allows a reasonably quick way to gather up whatever process documentation exists in your infrastructure, select the ones you want to keep and tie them together in a logical fashion. It avoids extra labor needed to reformat documents or efforts to "document everything".

Process Guides, documents that describe each ITSM process you are using can be used as the vehicle for describing activities and procedures that part of a process. Procedure steps show links to existing work instruction documents or items within your infrastructure.

The following example illustrates how all of this can come together.

In the example below, a procedure has been developed using a flowchart style to describe the workflow for executing a process activity. Flowchart boxes are linked to detailed work instructions where necessary.

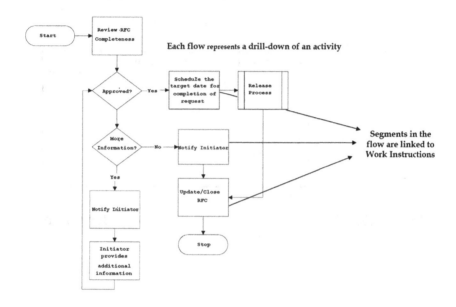

Figure 21: Flowchart Example

Detailed work instructions provide the lowest level descriptions of how a procedure might be performed. Work instructions may exist all over your infrastructure, or can be developed separately.

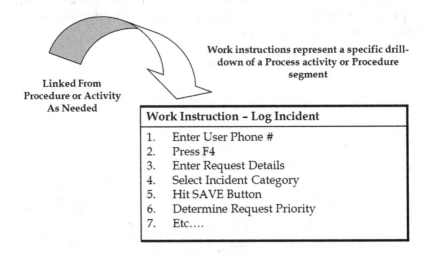

Linked From
Procedure or Activity
As Needed

Work instructions represent a specific drill-down of a Process activity or Procedure segment

Work Instruction – Log Incident
1. Enter User Phone #
2. Press F4
3. Enter Request Details
4. Select Incident Category
5. Hit SAVE Button
6. Determine Request Priority
7. Etc....

Figure 22: Work Instructions Linked To Procedure Steps

Within the process guide, an additional table can be added for each procedure that has links to more detailed work instructions. That table might look like the following:

Procedure—ACCEPT CHANGE		
Step	Description	Example Work Instruction Link
1	Review RFC for completeness	http://url_for_RFC-review.doc
2	Approved?	
3	Schedule target date for completion of the request	See vendor manual page 87 for request scheduling
4	Pending more information?	
5	Notify initiator	http://url_for_escalation_list.xls
6	Initiator provides additional information	http://url_for_RFC-form.doc
7	Notify initiator	http://url_for_escalation_list.xls
8	Update and close RFC	See vendor manual page 122 for how to close an RFC

Note that the work procedure links act as the vehicle for identifying disparate items in the infrastructure. Each procedure box is referenced and an associated link is provided where appropriate.

Not every procedure step needs to specify a work instruction—only those where deemed appropriate. You can also create new work instruction artifacts if they are needed and nothing else exists.

Chapter

7

Service Architecture

Service Architecture Overview

This chapter provides a service view of the IT Service Management architecture. The service view presented here consists of service lines that have been assembled into the following model:

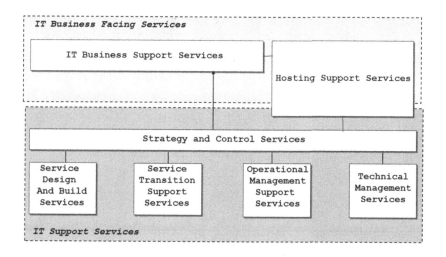

Figure 23: Service Architecture

In the above model, the service lines have been organized into IT Business Facing services and IT Support services. The business facing services are the ones most typically seen by the business. The IT services are more in the background and seen mostly by IT operations, support and development units.

The Service Lines are loosely based on the ITSM Service Lifecycle. Subsequent sections of this chapter describe the kinds of services that are included in each service line.

It is the author's view that IT Support Services, that is, those below the dotted line in the model are the same regardless of IT organization or company you work for. IT Business Facing services, however, will be different based on your company, market share and industry.

Detailed descriptions of each service described in this model as well as features of those services can be found in the Servicing ITSM book.

Technical Management Services

This Service Line includes services that implement, operate and maintain types of physical IT service assets that underpin IT and business services. Services in this line are grouped by technology platforms. A recommended set of services for this line are as follows:

- Server Management
- Database Management
- Application Management
- Network Management
- Storage Management
- Print Management
- Fax Management
- Physical Facilities Management
- Telephony Management
- Personal Computing Device Management
- Mobile Device Management
- Specialized Device Management
- Virtualization Management
- IT Supplies Management
- Middleware Transaction/Message Management

These kinds of services are probably the most familiar to IT technicians and support staff. They are described in much more detail in the Servicing ITSM book.

Operational Management Support Services

This Service Line includes services that manage and maintain IT operational workflows that cut across all the IT service assets in the physical infrastructure. Services in this line are grouped by those operational workflows that are necessary, but not directly supportive of business functions. A recommended set of services for this line are as follows:

- Service Desk
- Service Monitoring
- Incident Response
- Problem Control
- Request Fulfillment
- Backup/Restore Management
- Job Schedule Management
- Dispatch and Break-Fix Support
- Clock Management
- Service Startup/Shutdown Management
- File Transfer and Control Management
- Archive Management
- Data Entry Support
- Report Packaging and Distribution Support

These kinds of services are usually more familiar to IT operators and operational support staff. They provide service support needed to keep services running day-to-day. They are described in much more detail in the Servicing ITSM book.

Service Transition Support Services

This Service Line includes services that provide support for transitioning new or changed service solutions to production operations. These services take a service solution or a service change and move it from a development state into an operational state. A recommended set of services for this line are as follows:

- Release Planning and Packaging
- Service Deployment and Decommission
- Site Preparation Support
- Service Validation and Testing Support
- Training Support
- Organizational Change Support
- Knowledge Management

These kinds of services are somewhat familiar to IT developers and operational support staff. They land somewhere in the middle between development and operations so service ownership and responsibility may be new to many IT organizations who typically fragment these services among many groups. They are described in much more detail later in the Servicing ITSM book.

Solution Design and Build Services

This Service Line includes services that plan, build and construct new services or changes to existing services. They take requirements for new services or service changes and translate these into service solutions. A recommended set of services for this line are as follows:

- Operational Planning and Consulting
- Solution Planning and Development
- Development Support Operations
- Capacity Management
- Availability Management
- Service Continuity Management
- Website Support

These kinds of services are generally familiar to IT developers and technical support staff. They also land somewhere in the middle between development and operations so service ownership and responsibility will also be somewhat new to many IT organizations. They are described in much more detail in the Servicing ITSM book.

Strategy and Control Services

This Service Line includes services that govern and manage service delivery quality. It includes services that provide strategies for how services will be delivered. It also provides controls around their delivery to ensure that services are delivered safely and in compliance with industry and government regulations. A recommended set of services for this line are as follows:

- IT Service Strategy Support
- Architecture Management and Research
- IT Financial Management
- IT Project Management
- Change Control
- Configuration and Asset Management
- Lease and License Management
- Access and Security Management
- Service Audit and Reporting
- IT Workforce Management
- Procurement Support
- Process Management
- Supplier Relationship Management

These kinds of services are also generally familiar to IT developers and operational support staff. They are described in much more detail in the Servicing ITSM book.

Hosting and Cloud Support Services

This Service Line includes bundles of IT support services that host IT functions. Services in this line can either serve customers directly or be used internally by IT support staff. Generally these kinds of services will use internal or external cloud-based delivery infrastructures. These might look like time sharing or outsourcing from a customer perspective. Examples of hosting support services could include:

- Basic Support Service
- Infrastructure As A Service (IAAS)
- Platform As A Service (PAAS)
- Application As A Service (AAAS)
- Software as a Service (SaaS)
- Network as a Service (NaaS)
- Security as a Service
- Storage as a Service
- Equipment as a Service
- Secure Controlled Infrastructure Facility (SCIF)

These kinds of services are also generally familiar to IT developers and operational support staff. They are described in much more detail in the Servicing ITSM book.

IT Business Support Services

This Service Line includes IT business facing services that directly support company business functions, business partners and customers. They represent IT services that customers and users directly interact with.

Business facing services can vary greatly from company to company and industry to industry. Some services are supporting IT shared capabilities that business organizations use. Others may be specific to support of business functions and outcomes specific to the organizations that run the business of the company. Examples of IT shared services that businesses, no matter which industry they serve, might directly use could include:

- Desktop Support
- Data Warehousing and Business Intelligence
- Internet Telephony Service
- Email and Messaging
- Service Introduction
- Other IT Business Facing Services

The services in this line will vary greatly from company to company. For this reason, a sampling of typical services by company industry is included in the following pages.

Examples of Other Business Support Services

This section provides a generalized starter set of possible business support services that might exist in your organization. It is based on industry.

General Corporate Services

- General Ledger
- Accounts Receivable
- Accounts Payable
- Annual Report
- Financial Reporting
- Audit Support
- Payroll
- Human Resources
- Tax Accounting and Reporting

Manufacturing Support Services

- Materials Management
- Receive Orders
- Fulfill Orders
- Shipping
- CAD Design Support
- Factory Floor Automation
- Manage Suppliers
- Supply Partner Relationship Management

Marketing Support Services

- Market Research
- Market Planning
- Product History Maintenance
- Competitive Analysis
- Customer Insight
- Market Segmentation

Sales Back Office Support Services

- Order Management
- Billing
- Collections
- Commissions
- Customer Profile Management

Sales Front Office Support Services

- Response Management
- Contact Management
- Lead Management
- Opportunity Management
- Sales Forecasting
- Channel Management
- Sale Management
- Pipeline Analysis

Customer Support Services

- Service Analytics
- Solution Administration
- Case Management
- Self Service
- Customer Call Center
- Customer Survey Support Services

Product Support Services

- Contract Management
- Warranty Management
- Professional Services
- Product Education
- Repair History
- Repair Dispatching
- Repair Scheduling
- Parts and Supply Inventory Management
- Product Planning
- Product Description
- Development Lifecycle
- Campaign Planning

Procurement Support Services

- Supplier Certification
- Contract Initiation
- Vendor Management
- Purchases
- Payments

Educational Support Services

- Enrollment
- Registration
- Student Loan Processing
- Tuition Management
- Scholarship Management
- Course Scheduling
- Classroom Scheduling
- Pay Teaching Staff
- Course Catalog
- Research Support
- Library Management
- Student Housing Support
- Classroom Infrastructure Support
- Online Course Delivery and Management

Hospital Support Services

- Admissions, Discharges and Transfers
- Lab Support
- Medical Records
- Pharmacy
- Radiology
- Emergency Room Scheduling
- Medical Staff Scheduling
- Patient Care Support
- Patient Billing
- Medical Supplies Management

Energy and Utilities Support Services

- Customer Billing
- Order Fulfillment
- Meter Reading
- Infrastructure Repair and Management
- Power Generation
- Power Transmission
- Power Retailing
- Energy Development
- Environmental and Safety Support
- Load Profiling
- Fleet Maintenance
- Regulatory Control Support

Financial Trading Support Services

- Portfolio Tracking and Management
- Technical Analysis
- Fundamental Analysis
- Trading Support
- Trading Media Management
- Risk Management
- Currency Management
- News Feeds and Distribution

Insurance Support Services

- Agent Commissions and Pay
- Claims Payment
- Policy Underwriting
- Policy Sales Support
- Fraud Detection Support
- Property Risk Management
- Catastrophe Risk Management and Reporting
- Community Program Support

Patent and Trademark Support Services

- Patent Search
- Patent Applications Management
- Maintenance Fee Management
- Patent Publishing
- Patent Document Maintenance

Banking Support Services

- Automated Teller Support
- Online Banking
- Account Management
- Retail Banking Support
- Commercial Banking Support
- Mortgage Application Support
- Credit Card Operations
- Regulatory Management and Reporting
- Credit Risk Management
- Teller Support

Determining Other Business Support Services

Other Business Support Services is the one category that will differ greatly from one business to the next. These services support the lifeline for your business organization such as sales support, manufacturing or accounting.

The challenge is to determine what these actually are. IT will typically make broad assumptions or guess at what these are. They may also end up confusing services with features of a service, how it is delivered, or the functional organization it is delivered from. Before you know it, a massive list of business support services has been compiled that far exceeds the IT organization's ability to effectively manage them.

Here is an approach that can be used to better determine what the Other Business Support Services actually are. This approach works well in accurately identifying what these services consist of and can quickly knock down sizeable application inventories to a more manageable number of services.

To start, collect the following information:

- ✓ Listings of application inventories
- ✓ IT Service Continuity Plans
- ✓ Company organization charts

Pull a small team together to sort through these with a series of working sessions. The team should be composed of representatives from IT operations and development units. IT service liaisons are also quite helpful as well as some representation from the business itself.

During the working sessions, walk through the inventories line by line. For Application A, for example, what business process, service or function does it support? Develop a service name for it. For example, if Application A supports the company Sales functions, then you might call the service something like Sales Support.

Now look at the next application. Would it also fall under the Sales Support service? The criteria for this might be that the application performs similar functions or is used by the same business unit. If not, create another service that it should fall under. Repeat this step for each application you come across in your inventory.

Note that for each item in your application inventory, some will represent new services. The vast majority, however, will merely represent a feature of an existing service.

IT Service Continuity Plans are also a good source. These will indicate a more enterprise perspective of what the company sees as their services. It also points out which services are considered critical to company business operations.

Company organization charts are also helpful in that they provide clues as to what the business feels their business services really are. These would help point out business service areas such as Sales, Marketing, Finance or Manufacturing. Identified areas could then be translated into IT Business Support services such as Sales Support, Marketing Support, Finance Support and Manufacturing Support services.

As you go through the inventory, it is recommended that you track the results of these activities into a spreadsheet. At a minimum, this spreadsheet would have one column with the Service Name and one column with the name of an application associated with it. An example might look like the following:

Application	Service
AP Reimbursibles	Finance Accounting and Support Services
Facsys/FileNET Print Admin	Finance Accounting and Support Services
Fax Sr.	Finance Accounting and Support Services
FileNet	Finance Accounting and Support Services
Filenet AP Invoice Imaging	Finance Accounting and Support Services
SAP TEIC	Finance Accounting and Support Services
Accounts Recievable Sundry	Finance Accounting and Support Services
SAP FI-AP Accounts Payable	Finance Accounting and Support Services
SAP FI-AR Accounts Receivable	Finance Accounting and Support Services
Mercury Diagnostics	Application Planning and Development Services
Mercury Performance Center	Application Planning and Development Services
Mercury Quality Center	Application Planning and Development Services
PI Development Server	Application Planning and Development Services
Mantis Bug Tracking System	Application Planning and Development Services
Clarity 8.0	Application Planning and Development Services
Time Reporting System	Application Planning and Development Services
Time Sheet Professional	Application Planning and Development Services
SIR Workbench	Application Planning and Development Services
ASR Application System Request	Application Planning and Development Services
AllFusion Endevor ACM (z)	Application Planning and Development Services
AllFusion Endevor Change Manager (z)	Application Planning and Development Services

Figure 24: Associating Applications with Services Example

Do not be surprised if you find this to be a long list. There may be a lot of applications to go through. For those services that consist of many applications, you may wish to add a third column that identifies a brief sentence about what that application does. This will translate into a service feature for the service that the application falls under.

Using the example above, the Time Sheet Professional application would represent a feature of the Application Planning and Development Service which is ". . . provides easy time and labor entry for developers on development projects . . .".

Once completed, look at the number of services that were identified. Ideally, this number should be less than 100. If you have more than that, it means you may be working at too fine a level of detail. Keep in mind that the more services you define, the more overhead and administration you will create in defining, managing, operating and reporting on services in your infrastructure.

Some companies like to brag how they operate with a list that numbers thousands of services. Observations have shown that these companies typically have a long list of services that no one pays much attention to. Try to keep your list as small as reasonably possible. A list of about 40-70 services for a mid-size to large business should be generally reasonable.

What happens if your spreadsheet shows a large number of services? Take a second pass through it with your team and look for opportunities to combine services. For example, you may have a list that shows support services for General Ledger, Accounts Receivable, and Accounts Payable. Consider combining all those services into a single Corporate Accounting Support service.

Chapter

8

Assessing the Tooling Architecture

Tool Architecture Assessment Overview

When looking at the tooling strategy, the current state of the support technologies is examined to determine how well they support IT Service Management activities and practices.

Typical areas of concern to look for when assessing tools are as follows:

- Incomplete support of the functional architecture—in other words—missing tools to support key activities
- Tools that have not been implemented to their best advantage
- Utilization of many point tools that are not well integrated
- Tools that create more problems than they solve
- Use of redundant tools to do the same things (i.e. more than one Incident Management tool)
- Tools and technologies that have no vendor support, custom built or out of warranty

The outcome for a tooling assessment should yield recommendations for tool gaps that need to be overcome. These then translate into tooling improvement projects and possibly changes in toolsets or additional tools.

A suggested set of assessment criteria for reviewing tools is shown on the following pages. Note that each area can be scored as suggested. The higher the score, the more value or effectiveness of the tool exists for supporting IT Service Management activities and processes.

A general meaning for each score level can be described as follows:

Score	Level	Meaning
0	None	Tool should probably be scrapped or replaced
1	Weak	Little or no value coming from tool—should probably be scrapped or replaced
2	Fair	Provides some value, but may wish to review opportunities for replacement or building capabilities around the tool
3	Capable	Provides adequate support for processes but should be reviewed for adequacy on a periodic basis
4	Mature	Meets all current needs but should be reviewed for adequacy on a periodic basis
5	World Class	Provides valuable support for process activities within the IT infrastructure

Tooling Functionality Assessment

This assessment determines how well a given tool supports IT Service Management process activities.

Criteria	Score
Provides no support for any IT Service Management activity—may not be used	0
Provides little support for IT Service Management processes	1
Provides support that applies to some activities of an IT Service Management process, but not all of the process	2
Provides support that address most activities of an IT Service Management process	3
Provides complete support for an IT Service Management process covering all of its activities	4
Provides complete support for an IT Service Management process covering all of its activities with integration to other processes	5

Tooling Platform Assessment

This assessment determines how well a given tool supports technology platforms in the IT infrastructure.

Criteria	Score
Provides no support for any technology platforms used in the IT infrastructure	0
Only provides support for a single technology platform but not fully	1
Only provides support for a single technology platform	2
Provides support for more than one technology platform	3
Provides support for all platforms included within the scope of a process	4
Provides support for all platforms within the scope of the entire IT infrastructure	5

Tooling Integration Assessment

This assessment determines how well a given tool integrates with other tools and technologies in the IT infrastructure.

Criteria	Score
Tool does not integrate with any other tools	0
Tool integrates with other tools but only via flat files, text files or other batch interfaces.	1
Tool provides a common user interface and integrates with flat files or other batch interfaces	2
Tool provides a common user interface and integrates through common API programming interfaces and data sharing	3
Tool shares data and cooperates with other tools to achieve management tasks transparently across different platforms	4
Tool shares data and cooperates with other tools to achieve management tasks transparently across all platforms in the IT infrastructure	5

Tooling Automation Assessment

This assessment determines how well a given tool automates IT Service Management process activities.

Criteria	Score
Use of tool requires more manual labor to work with than if the tool were not in place	0
Use of tool requires high levels of manual labor to work with	1
A significant proportion of process activities are manually initiated, with little automated functionality provided	2
Automation is reactive and works with operator direction	3
Proactive features with considerable automation but exception conditions need to be handled manually	4
High degree of automation provided by the tool with predictive and self-healing capabilities that require almost no human intervention	5

Tooling Usability Assessment

This assessment determines how easy a given tool is to use by IT service support and delivery staff.

Criteria	Score
Use of the tool results in incidents, inaccurate data or viewed as unusable	0
Tool is viewed as highly confusing and hard to use	1
Usable only by the most highly skilled technical and senior staff	2
Basic functions of the tool easy to use; advanced usage requires high skill level	3
Most tool functions can be effectively used by those with low technical skill levels	4
All tool functions can be used effectively by those with low technical skill levels	5

Tooling Reporting Assessment

This assessment determines how the level of reporting capabilities for a given tool.

Criteria	Score
Tool provides little or no reporting capabilities	0
Tool only provides a limited set of standard reports	1
Tool only provides a limited set of standard reports with minor ad-hoc reporting capabilities	2
Tool provides a set of standard reports— customized and ad-hoc reports can be produced but only with the aid of experienced users or technical support personnel	3
All reports available with the tool can be customized with ad-hoc reporting available for all users	4
Tool provides easy customization and scheduling of all needed reports for a process including ad hoc reports that can be initiated automatically as well as manually with automated exception reporting	5

Tooling Data Assessment

This assessment determines how easy a given tool is to use by IT service support and delivery staff.

Criteria	Score
Tool provides little or no data	0
Tool produces limited data to flat text file	1
Tool data is written to a file, but data from multiple runs cannot be combined easily and there is no search capability	2
Tool data is added to a cumulative database and search on single keys is possible	3
Tool data is kept in a single database with easy searching for specific data	4
Tool data is kept in a database along with data from other management tools and is easy to find and combine with other tools, time frames, or other factors	5

Tooling Communication Assessment

This assessment determines how well a tool communicates events and status information throughout the IT infrastructure.

Criteria	Score
Tool provides no communication events	0
Tool displays results on a console terminal only where those results would need to be manually viewed	1
Tool interfaces to only a few communication vehicles such as pagers, cellular phones, E-mail but requires manual initiation	2
Tool interfaces to multiple varieties of some communication vehicles	3
Tool interfaces to most communication vehicles	4
Tool interfaces with all standard communication vehicles automatically	5

Tooling Openness Assessment

This assessment determines how open a tool is in terms of its ability to operate on technology platforms in the infrastructure and compatibility with industry standards.

Criteria	Score
Tool does not work with any technology platforms in the IT infrastructure	0
Tool only runs on a single platform	1
Tool runs in a single environment, but has operates with a proprietary interface	2
Tool is adaptable to several environments or works with defined standard interfaces	3
Tool is adaptable to installed technology and desired standard interfaces	4
Tool is adaptable to a wide variety of technology	5

Support Assessment

This assessment determines how well a tool is supported by third party vendors.

Criteria	Score
Tool is completely built and maintained with internal IT staff	0
Tool support is provided only through specialized consultants not part of the tool vendor's organization or the vendor has declared the tool as non-strategic	1
Tool support is available on a limited basis or the vendor has warned that the tool will not be supported in a future timeframe	2
Tool solution is fully supported by the vendor but support is only provided during standard business hours	3
Tool solution is fully supported by the vendor at any time or day	4
Tool solution is fully provided by the vendor at any time or day and the vendor provides a variety of self-help, remote diagnostic facilities and training	5

Mapping Tools with Processes

In order to assess completeness of the entire tooling set, the entire tool inventory should be mapped against the complete set of IT Service Management Processes. This can be done in a manner as shown in the following example:

Tool	Vendor	Platform						Processes			
		Facilities	Network	Hardware	Systems Software	Applications	Storage	Incident Mgt	Problem Mgt	Change Mgt	etc...
Tool 1	Vendor 1		X					X	X		
Tool 2	Vendor 1	X	X	X	X	X	X			X	
Tool 3	Vendor 2					X		X	X		

Figure 25: Mapping Tools With Processes They Support Spreadsheet Example

The purpose of this assessment is to see if there are process areas that may be lacking tool support. It can also indicate whether redundant tool sets are in place or whether platforms are missing management tools.

As an option, the X's in the above diagram can be replaced with a number that represents the average score of the assessments described earlier in this chapter.

Even further, the process cells may be color coded based on that score. For instance, scores 1-2 may be colored red, 3-4 yellow and 5-6 green. This presents a nice visual heat map of how well the tools are supporting processes used in the IT infrastructure.

The download website with this book includes an Architectural Analysis tool that can assist with the mapping techniques shown above.

Mapping Tools with the Functional Architecture

Another technique to assess completeness of the entire tooling set is to map the current tool set to the entire functional architecture presented in this book. The approach would be similar as against the processes, but might look like the following:

Tool	Vendor	Platform						Capabilities			
		Facilities	Network	Hardware	Systems Software	Applications	Storage	Access Mgt System	ACD System	Asset Mgt System	etc...
Tool 1	Vendor 1		X	X	X		X			X	
Tool 2	Vendor 1	X	X	X	X	X	X	X			
Tool 3	Vendor 2					X			X		

Figure 26: Mapping Tools with the ITSM Architecture Spreadsheet Example

The purpose of this assessment is to see if there are missing functional areas not addressed by any tools. This kind of assessment may be especially helpful when building new data centers or processing facilities to ensure that a complete tooling solution has been put into place.

The download website with this book includes an Architectural Analysis tool that can assist with the mapping techniques shown above.

Chapter
9

Architecture Governance

Architecture Governance Overview

Architecture Governance is about managing the design, deployment, maintenance, and evolution of the IT Service Management enterprise architecture and ensures an ongoing service improvement lifecycle is maintained. To accomplish this, Architecture Governance will work in partnership with key stakeholders and the ITSM Steering Committee to continually align the architecture with business and service strategies.

Employment of an Architecture Governance solution should not be overlooked. It plays a vital role in:

- Meeting the goals of creating global process standards with approved local variations where needed to meet business objectives.
- Providing an independent channel and single point of contact for resolving ITSM architecture conflicts.
- Managing key architecture changes over time to meet the needs of the business

- Coordinating research into new technologies and best practices for IT Service Management that may provide value to the business organization.

A number of guiding principles can be used when putting the Architecture Governance solution together. These include principles such as the following:

- The ITSM Steering Committee will be the final decision making authority for ITSM architecture changes and will own the governance process.
- An exception and appeals process will be in place to resolve ITSM architecture issues and concerns
- Implementation of ITSM architecture changes will utilize the ITSM Continual Service Improvement model.

Four main sub-processes make up the ITSM Architecture Governance solution. These are summarized as follows:

Architecture Governance

This sub-process provides a structured approach for reviewing and approving decisions for architecture changes to be made in accordance with company standard IT management solutions.

Architecture Exceptions and Appeals

This sub-process provides a means of escalating ITSM decisions for the use of non-conforming solutions to meet unique and/or local business requirements.

Architecture Vitality

This sub-process provides a way to incorporate new ITSM solutions and changes into the enterprise architecture standard as a result of changing business needs. This sub-process also researches new ITSM solutions and best practices in the marketplace.

Architecture Communications

This sub-process provides a means for syndicating ITSM architecture solutions as they evolve across the business enterprise.

An illustration of how these fit into an overall ITSM Architecture Governance Model can be shown as follows:

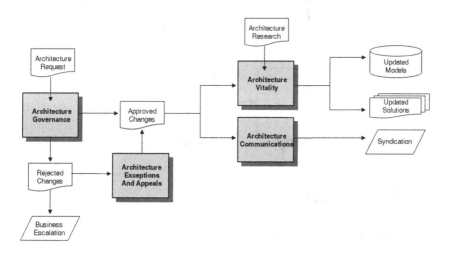

Figure 27: ITSM Architecture Governance Model

Architecture Governance Workflow

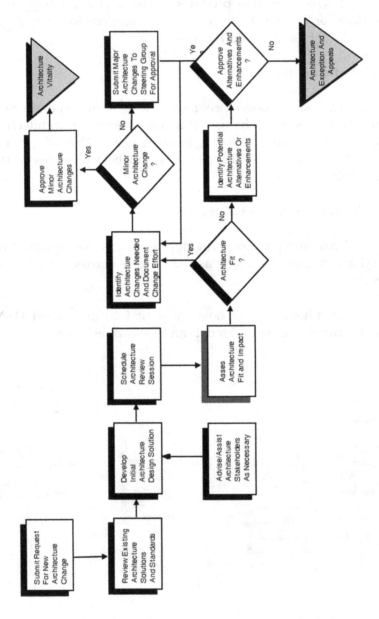

Figure 28: Architecture Governance Procedure

Architecture Exception & Appeals Workflow

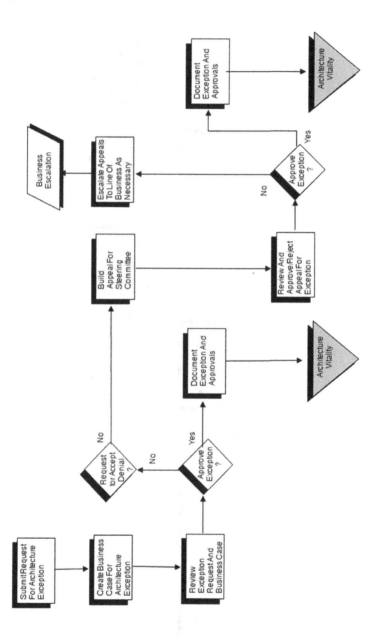

Figure 29: Architecture Exception and Appeals Procedure

Architecture Vitality Workflow

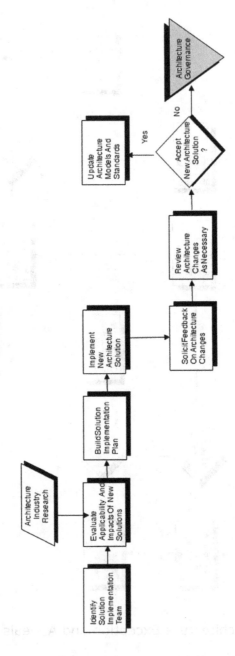

Figure 30: Architecture Vitality Procedure

Architecture Communications Workflow

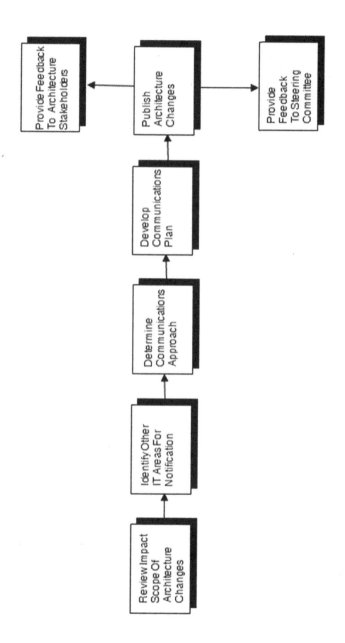

Figure 31: Architecture Communications Procedure

About the Author

Randy A. Steinberg has extensive IT Service Management and operations experience gained from many clients around the world. He authored the ITIL 2011 Service Operation book published worldwide. Passionate about game changing management practices within the IT industry, Randy is a hands-on IT Service Management expert helping IT organizations transform their IT infrastructure management strategies and operational practices to meet today's IT challenges.

Randy has served in IT leadership roles across many large government, health, financial, manufacturing and consulting firms including a role as Global Head of IT Service Management for a worldwide media company with 176 operating centers around the globe. He implemented solutions for one company that went on to win a Malcolm Baldrige award for their IT service quality. He continually shares his expertise across the global IT community frequently speaking and consulting with many IT technology and business organizations to improve their service delivery and operations management practices.

Randy can be reached at

RandyASteinberg@gmail.com.